# Current Topics in Microbiology 243 and Immunology

Editors

Springer
Berlin
Heidelberg
New York
Barcelona
Hong Kong
London
Milan
Paris
Singapore
Tokyo

# Combinatorial Chemistry in Biology

Edited by M. Famulok,
E.-L. Winnacker, and C.-H. Wong

With 48 Figures and 8 Tables

Springer

Professor Dr. MICHAEL FAMULOK
Kekule-Institut für Organische Chemie
und Biochemie
Gerhard-Domagk-Str. 1
D-53121 Bonn
Germany

Professor Dr. ERNST-L. WINNACKER
Ludwig Maximilians Universität München
Institut für Biochemie – Genzentrum
Feodor-Lynnen-Str. 25
D-81377 München
Germany

Professor Dr. CHI-HUEY WONG
The Scripps Research Institute
10550 North Torrey Pines Road
La Jolla, CA 92037
USA

*Cover Illustration: "The cover shows a chalk-drawing on a blackboard symbolizing the principle of randomization and selection by successively selecting patterns on the face of dices. The cover was generated by Boris Steipe and Iris Häussler, BioLink GmbH, Martinsried."*

*Cover Design: design & production* GmbH, Heidelberg

ISSN 0070-217X
ISBN 3-540-65704-5 Springer-Verlag Berlin Heidelberg New York


Library of Congress Catalog Card Number 15-12910
Printed in Germany



Typesetting: Scientific Publishing Services (P) Ltd, Madras

Production Editor: Angélique Gcouta

SPIN: 10679885 27/3020 – 5 4 3 2 1 0 – Printed on acid-free paper

# Preface

The essence of combinatorial chemistry or techniques involving "molecular diversity" is to generate enormous populations of molecules and to exploit appropriate screening techniques to isolate active components contained in these libraries. This idea has been the focus of research both in academia and in the pharmaceutical or biotechnology industry. Its developments go hand in hand with an exploding number of potential drug targets emerging from genomics and proteomics research.

When the editors of *Current Topics in Microbiology and Immunology* encouraged us to assemble the present volume on *Combinatorial Chemistry in Biology*, we immediately felt that this might prove quite beneficial for the audience of this series. The field of combinatorial chemistry extends over a broad range of disciplines, from synthetic organic chemistry to biochemistry, from material sciences to cell biology. Each of these fields may have its own view on this topic, something which is reflected in a growing number of monographs and "special editions" of journals devoted to this issue or aspects thereof. The title of the present volume of Springer-Verlag's series suggests that it also has its own special focus. And, generally speaking, this is not wrong: we would even claim the special focus of this volume is on the immunologically relevant aspects of combinatorial chemistry. However, we thought it might also be quite appropriate to include some chapters from areas normally not immediately associated with immunology, in order to attract the attention of researchers working in different fields and maybe even "seed" some new ideas.

In the first and most detailed part of volume, an overview of various strategies of encoded combinatorial chemistry of peptides or proteins is presented. These techniques are represented by peptide, phage-display, ribosome-display, RNA/peptide-fusion, and spatially oriented peptide libraries, as well as libraries of proteins expressed by cells – the various two-hybrid technologies. This section also contains a chapter on evolutionary aspects of protein engineering. We then move into the world of nucleic acid libraries and their screening for functional molecules such as "aptamers" – specific ligand-binding nucleic acids – and ribozymes – nucleic acids which are catalytically active. This is followed by a chapter on dynamic combinatorial chemistry of libraries of small synthetic organic molecules. Finally, the volume closes with a chapter on the perspectives of "synthetic biology" and its relevance for evolutionary biotechnology.

We would like to thank all contributors for their efforts, the editors of the *Current Topics* series, and the staff at Springer-Verlag for their continuous help and support in getting this volume published.

Michael Famulok
Chi-Huey Wong
Ernst-Ludwig Winnacker

# List of Contents

# List of Contributors

(Their addresses can be found at the beginning of their respective chapters.)

# Peptide Libraries in Cellular Immune Recognition

S. Uebel[1], K.H. Wiesmüller[2], G. Jung[3], and R. Tampé[1]



## 1 Principles of Cellular Immune Recognition

Protective immunity against microbial pathogens is connected with the ability to discriminate between self and non-self. Vertebrates have evolved an adaptive immune system in which specialized white blood cells, the T cells, fulfill this role. These cells are trained to recognize proteolytic fragments of viral, bacterial or parasitic antigens in the context of specialized antigen presenting proteins, the major histocompatibility complex (MHC) molecules. Recognition of a non-self peptide in complex with the self component, the MHC molecule, ultimately leads to destruction of the infected cell or to the production of neutralizing antibodies.

Cytotoxic T cells constantly monitor the status of a cell via the set of peptides found on the surface of this cell in association with MHC class I molecules (Townsend and Bodmer 1989; Yewdell and Bennink 1992; Rammensee et al.

[1] Department of Cellular Biochemistry and Biophysics, Institute for Biochemistry, Philipps-University Marburg, Medical School, Karl-von-Frisch-Str. 1, D-35033 Marburg, Germany
[2] EMC microcollections, Sindelfinger Str. 3, D-72070 Tübingen, Germany
[3] Institut für Organische Chemie, Universität Tübingen, Auf der Morgenstelle 18, D-72076 Tübingen, Germany

1993). These peptides are derived from intracellular proteins targeted for degradation and are processed as outlined below (Fig. 1). In the case of a productive infection during which microbial proteins are produced, non-self peptides are found on MHC class I molecules in addition to self peptides. Cytotoxic T cells are trained in the thymus in a sequence of positive and negative selection events to recognize peptide-loaded MHC class I molecules via the T cell receptor and to destroy cells showing non-self peptides.

Extracellular antigens instead are subject to binding of neutralizing antibodies, triggering ingestion by phagocytotic cells, or destruction of antigen-bearing bacteria through the complement system. Production of antibodies requires processing of the extracellular antigens by specialized antigen-presenting cells, including bone-marrow derived B cells, and presentation of peptides in association with MHC class II molecules (Fig. 1). Non-self peptides are recognized by T helper cells, leading to stimulation and clonal expansion of antibody-producing B cells bearing the right specificity (Sant and Miller 1994).

## 1.1 Processing and Presentation of Antigens – A Short Overview

Removal of damaged, misfolded or otherwise unnecessary proteins is a function important for the survival of a cell. Proteins marked for degradation by ubiquitinylation are proteolytically processed by the proteasome complex (Rock et al. 1994). From this pathway, peptides 8–16 amino acids in length are diverted for binding to MHC class I molecules after translocation into the endoplasmic reticulum (ER) by the transporter associated with antigen processing (TAP), localized in the ER-membrane (Hill and Ploegh 1995; Koopmann et al. 1997). Here, assembly of class I and $\beta_2$-microglobulin and subsequent peptide loading takes place. Fully assembled complexes are routed to the cell surface for T cell recognition in the exocytotic pathway.

MHC class II molecules instead are loaded with peptides generated in the endocytotic pathway. Extracellular antigen is taken up by antigen presenting cells (APCs) through receptor-mediated endocytosis, phagocytosis or other endocytotic events and is sequentially degraded as it travels from early to late endosomes and ultimately to lysosomes (Fig. 1) (Gosselin et al. 1992; Vidard et al. 1992). Empty class II molecules on their exocytotic route cross this degradation pathway and are loaded with partially degraded antigen in the MHC class II-containing compartment MIIC, resembling late endosomes or early lysosomes (Pieters et al. 1991). Upon loading they leave to the cell surface, again for T cell recognition, while remaining antigen is further degraded to amino acids.

## 1.2 MHC Molecules – The Common Principle

The similarity of MHC class I and II structures reflects the common function, namely binding of peptides while allowing access for the T cell receptor. In order to

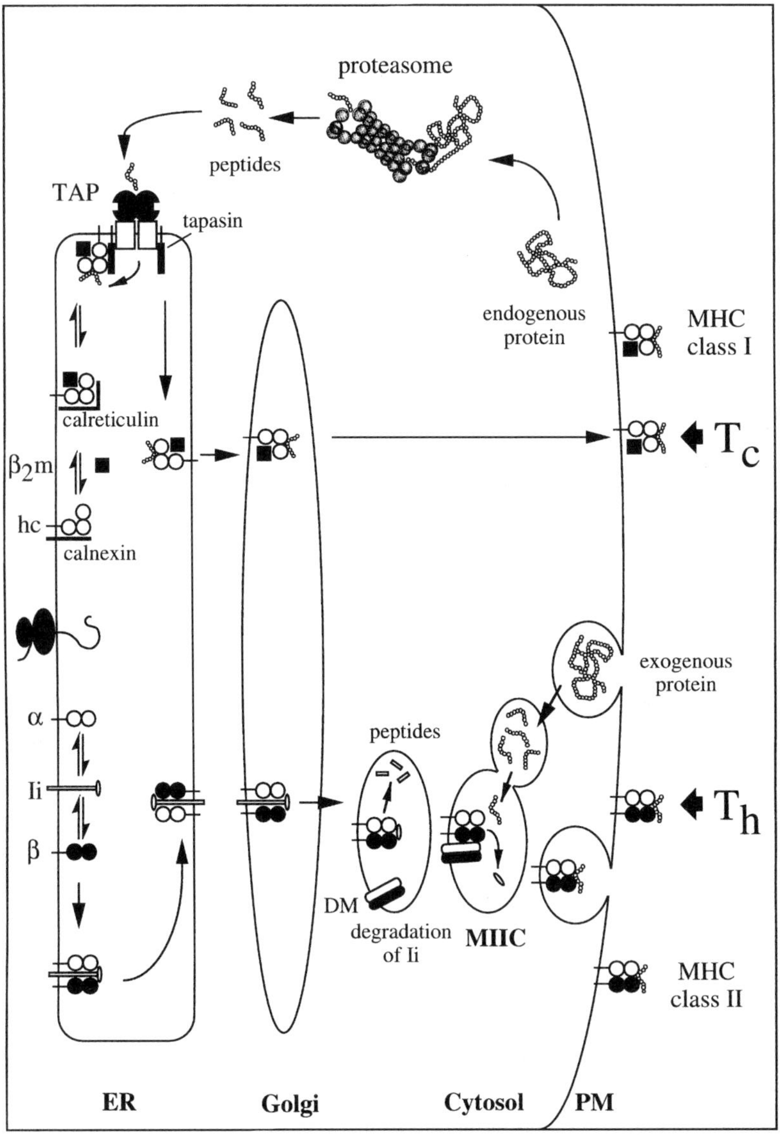


**Fig. 1.** Antigen processing and presentation via MHC class I and II molecules. Endogenous proteins are degraded in the ubiquitin-proteasome pathway. The generated peptides can be transported into the endoplasmic reticulum by the transporter associated with antigen processing (TAP), where they are loaded onto class I molecules in a tightly regulated process involving the chaperones calnexin, calreticulin and tapasin. Fully assembled class I-peptide complexes subsequently dissociate from tapasin and traffic to the cell surface for recognition by cytotoxic T cells. Exogenous proteins in contrast are degraded as they travel down the endocytotic pathway. Loading of peptides occurs in the MHC class II containing compartment MIIC after degradation of the invariant chain and removal of CLIP by DM. Again, only fully assembled complexes exit to the cell surface for recognition by T helper cells. α,β, class II α and β chain; $\beta_2m$, $\beta_2$-microglobulin; *DM*, HLA-DM; *ER*, endoplasmic reticulum; *hc*, class I heavy chain; *Ii*, invariant chain; *PM*, plasma membrane; *Tc*, cytotoxic T cell; *Th*, T helper cell

maximize the number of non-self peptides that can be presented, each MHC molecule must be able to bind a wide spectrum of different peptides but with high affinity, matching the low peptide concentrations in the respective loading compartments. The structural organization of the MHC molecules places them in the "immunoglobulin fold" superfamily of proteins engaged in cell-cell contacts. The peptides are bound in a cleft formed by a β-sheet platform and two α-helices. In the case of class I molecules, the α1 and α2 subunits of the heavy chain form this peptide binding groove, with closed ends suited for peptides ideally 8–11 amino acids in length (Bjorkman et al. 1987). The cleft formed by the α2 and β2 domains of the class II chains, in contrast, has open ends, allowing for longer peptides to stick out at both ends (Brown et al. 1993). The principles underlying stable peptide binding to class I molecules (see below) lead to significant restrictions on the peptide repertoire that each class I molecule can bind. Thus, for presentation of a maximum number of different antigens, MHC molecules have to be highly polymorphic. Each human carries up to six of several hundred alleles of class I and class II molecules. This polymorphism plays a key role in allograft rejection reactions and has ultimately led to the identification of class I and II molecules as factors involved in histocompatibility. Also, allele-specific susceptibility for certain autoimmune diseases highlights the importance of understanding MHC–peptide interaction.

## 1.3 Assembly Pathways Are Different Between MHC Class I and II Molecules

In order to ensure proper folding, loading, and intracellular trafficking of MHC molecules, a variety of chaperones are involved (Fig. 1). Calnexin binds to the newly synthesized class I heavy chain in the ER (Degen and Williams 1991; Sugita and Brenner 1994) and is replaced by calreticulin after association with $\beta_2$-microglobulin ($\beta_2$m) (Sadasivan et al. 1996). Tapasin bridges empty heavy chain/$\beta_2$m complexes and TAP, possibly facilitating peptide loading (Sadasivan et al. 1996; Ortmann et al. 1997). Trimeric class I-$\beta_2$m-peptide complexes are then released from tapasin for further trafficking to the cell surface. The class II-loading pathway instead has to exclude premature association with ER-resident peptides, thus with peptides of intracellular origin. Therefore, class II molecules remain associated with the invariant chain Ii glycoprotein after synthesis in the ER until the class II-Ii complexes accumulate in endosomal compartments, with the CLIP region occupying the peptide binding groove of the class II molecules (Cresswell 1992). Upon entering endosomal compartments, Ii is sequentially degraded by proteases until only the CLIP fragment remains associated with class II. Another protein, the heterodimeric HLA-DM, then triggers removal of CLIP in the MHC class II containing compartment MIIC (Sherman et al. 1995). Class II molecules loaded with high affinity peptides then undergo a conformational change that leads to sorting to the cell surface.

### 1.4 T Cell Recognition

The heterodimeric T cell receptor (TCR) represents another member of the immunoglobulin superfamily. By genetic recombination, a total of more than $10^7$ different receptor molecules is generated, each expressed on a different cell. T cells undergo a series of selection events, in which cells with receptors reacting with self antigens and receptors not recognizing MHC structures are eliminated (KAPPLER et al. 1987). Mature T cells are mostly either CD4+ or CD8+, e.g. they carry the accessory molecules CD4, for class II recognition, or CD8, for the class I contact. Target cells, carrying non-self peptides in the MHC class I context, are triggered to undergo apoptosis following the release of lymphokines (regulatory proteins) and pore-forming enzymes by the CD8+ T cells, also called cytotoxic T cells (SCHWARTZ 1985). Instead, CD4+ or helper T cells recognizing non-self structures bound to class II molecules modulate the action of different immune effector cells, for example, antibody producing B cells, again by the release of lymphokines (SINGER and HODES 1983).

## 2 Combinatorial Libraries Applied in Studies with MHC Class I Molecules

Naturally occurring organic compound libraries are found within many structural classes of products from microorganism or plants (JUNG 1996). Natural peptide libraries in particular are involved in some of the most fundamental events in mammalian immune response and interact with receptors which are highly degenerate with respect to their peptide interactions. Natural peptide libraries have been characterized by pool sequencing (STEVANOVIĆ and JUNG 1993), and allele-specific sequence motifs for MHC class I and MHC class II molecules have been elucidated (FALK et al. 1991).

Complex mixtures of synthetic peptides were introduced in 1986 to identify artificial antigens. At selected coupling cycles during synthesis, more than one amino acid was incorporated, yielding a mixture on a solid support. Characteristic for different mixtures is the number and the position(s) of defined sequence positions at which only one amino acid was coupled (GEYSEN et al. 1986). Combinatorial synthetic peptide libraries have been introduced to search for an antigen within soluble hexapeptide libraries (HOUGHTEN et al. 1991; PINILLA et al. 1992) or to identify streptavidin binding peptides by screening pentapeptides immobilized on single resin beads (LAM et al. 1991). This initial studies have been extended to numerous applications in basic research and to identification of peptide lead structures for drug development (GALLOP et al. 1994). Several different strategies and formats have been described in combinatorial solid phase peptide synthesis. For example, soluble all D-amino acid peptide libraries have been used to define new ligands of opioid receptors with enhanced enzymatic stability in an iterative procedure characterized by decreased complexity in each successive step of screening and synthesis (Dooley et al. 1994).

Library formats were designed to ascertain multiple reuse of one type of library homogenous in length and end groups, and inhomogeneous in sequences and in the number and positions of defined residues. Other formats allow encoding of the synthesis history by oligonucleotides or chemical tags (Gallop et al. 1994). A commonly used procedure for the solid phase synthesis of, for example, tetrapeptide libraries is the split resin approach, which is, however, limited by the number of resin beads applicable in the synthesis (Furka et al. 1991). To synthesize highly diverse libraries carrying more than six degenerate sequence positions, several methods, using premixed mixtures of amino acids for the introduction of X-positions in the sequence, have been described (Wiesmüller et al. 1996). It is obvious that the composition of the amino acid mixtures and the chemistry applied for coupling as well as the nature of the solid support influence the distribution of individual sequences. This synthetic peptide libraries have been analyzed by electrospray mass spectrometry (Metzger et al. 1994) and pool sequencing (Stevanović and Jung 1993). Limitations of the application of peptide libraries are given by the stoichiometry. For example, combinatorial collections of $19^{15}$ peptides are only partially accessible for biological assays (Wiesmüller et al. 1996).

## 2.1 Binding Motif of MHC Class I Molecules

The rules for peptide selection by MHC class I molecules were determined: (1) by the characterization of peptide mixtures extracted from MHC complexes (Falk et al. 1991), (2) from the effects of collections of different peptides on binding to MHC molecules and from the response of cytotoxic T cells on peptides presented by MHC molecules (Chen et al. 1993), and (3) from crystal structure analyses of defined MHC-peptide and MHC-peptide-T cell receptor complexes (Madden et al. 1992; Fremont et al. 1992; Garboczi et al. 1996; Garcia et al. 1996). MHC class I ligands are frequently octa- or nonapeptides and obey allele-specific sequence motifs carrying prominent anchor residues (Rammensee et al. 1996). The peptide binding groove offers specific pockets for these anchor residues (Madden et al. 1992; Fremont et al. 1992).

In addition, complex synthetic peptide libraries have been used to further analyze the MHC-peptide interaction, especially to define the contributions of primary, secondary, and non-anchor residues in MHC-peptide binding and to investigate the response of cytotoxic T cells (Fig. 2). Some 152 octapeptide libraries $O/X_7$, each representing a collection of about $900 \times 10^6$ individual peptides, were investigated in initial studies with the mouse MHC class I molecules, H-2K$^b$ and H-2L$^d$. Thereby the contribution of individual amino acids in the sequence position O to the binding of peptides to MHC molecules was defined by a standard stabilization assay for in vitro detection of peptide-loaded MHC molecules. The results were not influenced by restrictions presented by intracellularly preselected, naturally processed peptides. Amino acids in positions O of the library $O/X_7$ were found

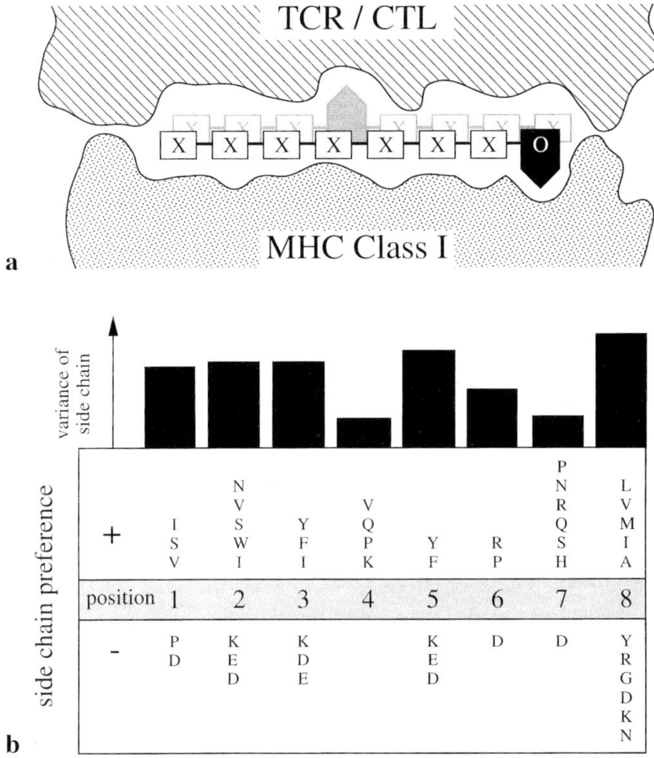


**Fig. 2.** **a** Example of the interaction of an octapeptide library with murine MHC class I molecules. The library is characterized by one prominent position (*black triangle*) for favorable interactions or unfavorable effects in contacts to MHC molecules or with the T cell receptor. **b** Ranking of amino acids according to the stabilization of mouse MHC H-2K$^b$ molecules by O/X$_7$ libraries (Udaka et al. 1995a). Amino acids O are classified by their influence on stabilization (+) or destabilization (–)

to contribute positively, negatively or neutrally to binding to the MHC molecule and the recognition of the resulting complexes by CTL (Udaka et al. 1995a,b; Pridzun et al. 1996). Using synthetic peptide libraries the determinative roles of motif amino acids, which are vital for peptide binding and for the stability of the entire complex, were confirmed.

Prominent primary anchor residues in anchor positions determined by pool sequencing (allele H-2K$^b$, positions 5 and 8; allele H-2L$^d$, positions 2 and 8) were ascertained by screening octapeptide libraries. Less important for stabilization are secondary anchor residues which were identified, e.g., for the H-2K$^b$ peptide complex in positions 1, 2, and 3 whereas amino acids in positions 4, 6 and 7 point out of the MHC binding groove and are accessible for interactions with the TCR. These positions are also characterized by a low number of residues unfavorable for stabilization of H-2K$^b$ (Fig. 2b). The contribution of individual amino acids on stabilization of MHC molecules that is induced by a ligand is not additive as influenced by flanking residues. Peptide libraries with two or more defined positions are in many cases highly sensitive to the structure and position of the second or further defined position. Peptides which lack one or both primary anchor positions also had significant effects on the stabilization mediated by secondary anchors (Udaka et al. 1995b). Unfavorable effects on binding can be induced by amino acids in non-anchor positions. Conformational analysis of several individual peptides bound to MHC class I molecules showed that the position of the peptide backbone in the binding groove as well as the orientation of side chains in other sequence positions is influenced by individual amino acid side chains (Madden et al. 1992; Fremont et al. 1992). These interactions strongly determine the accessibility of the peptide on the MHC I surface and thus the response of the effector cells. Anchor residues promote efficient binding to MHC I but notably suppress the CTL response to the complex peptide libraries (Udaka et al. 1995b).

Nonapeptide libraries were also used to determine the peptide binding motifs for three rat MHC class I molecules. Proper folding in the presence of the $X_9$ library, isolation of peptides from the MHC-peptide complex, and pool sequencing revealed a motif almost identical to the one obtained by classical procedures with acid-eluted peptides (Stevens et al. 1998). The motif obtained with the $X_9$ library is not strongly affected by cellular processing and transport and is thus similar to the stabilization experiments in the presence of $O/X_8$ libraries, a measure of the conformational space offered by the binding groove of the respective MHC molecule. The minor differences in the "synthetic" and "natural" motifs suggest that the proteolytic cleavage and TAP transporter specificities predetermine restrictions on the availability of peptides, which in turn is influenced by characteristic residue-sequence position combinations.

## 2.2 MHC Class I-Restricted T Cell Recognition

The degeneracy of peptide recognition by one TCR implies that many individual synthetic peptides can be used for priming of CTL with directed specificity. The induction of T cell-mediated immunity against tumor or viral antigens does not require identification of the natural epitopes; any peptide presented by MHC class I and recognized in the correct orientation by the TCR should be sufficient. Synthetic epitopes have been identified with the peptide library approach to determine the molecular basis for the observed cross-recognition of two ligands by a single receptor (Udaka et al. 1996). A prerequisite for the T cell-mediated immune response is the formation of the MHC-peptide complex and subsequent recognition by the T cell repertoire, which can be analyzed in cell lysis assays with $^{51}Cr$-loaded target cells. Chromium release is a measure for peptide-induced cell lysis by CTL and indicates the potency of the peptide preparation to serve as an epitope for the allele used in the assay.

The responses of a H-2K$^b$-restricted CTL clone specific for the octapeptide epitope SIINFEKL and the eight $O/X_7$ peptide libraries **S**XXXXXXX, X**I**XXXXXX, XX**I**XXXXX, XXX**N**XXXX, XXXX**F**XXX, XXXXX**E**XX, XXXXXX**K**X, XXXXXXX**L** that define the epitope are a measure of the contribution of the amino acids in the O-position (indicated in bold type) to peptide recognition by the CTL clone 4G3. Improvement in CTL response was observed with O in position 1, 2, 4, 6, and 7 of $O/X_7$ peptide libraries. Amino acids in MHC anchor positions 5 and 8 were extremely unfavorable for a CTL response (Udaka et al. 1996b). This observation is probably due to the competition induced by peptides that bind to the MHC molecule and do not present residues which evoke CTL responses. The unfavorable effects of anchor amino acids in position O of $O/X_7$ peptide libraries on CTL responses cannot be generalized. The screening of 152 $O/X_7$ peptide libraries on cytolysis of target cells by several CTL clones showed clone-specific activity patterns which allow the deduction of synthetic T cell epitopes effective in the concentration range also typical for naturally occurring epitopes (Gundlach et al. 1996a/b). The degree of degeneracy is characteristic for a certain CTL clone and is also represented in the activity pattern of the peptide library.

Peptide analogues can serve as synthetic epitopes which are functional mimics of the native sequence. Combinatorial peptide libraries with restricted numbers of building blocks in X-positions were screened using an iterative strategy based on their ability to sensitize RMA/S cells for lysis by CTL specific for recognition of unidentified tumor antigens. The approach succeeded in the construction of functional epitope mimics, provided that the presenting MHC molecule and the peptide-binding motif were characterized (Blake et al. 1996). One peptide was identified that induces a CTL response in vivo and thus represents the functional mimicries of a CTL epitope. The application of peptide libraries for the identification of synthetic epitopes for the potential application in tumor vaccines was practically confirmed by adoptive transfer of a CTL line from mice immunized with the epitope. The transferred CTL inhibited the growth of an ascitic tumor (Blake et al. 1996).

Combinatorial peptide libraries, either completely random or characterized by one or several defined positions, are useful tools for the identification of the critical features of MHC class I binding peptides and of natural and synthetic epitopes. The complete activity pattern of an $O/X_7$ library with 152 individual samples represents the influence of amino acid residues critical for mediating contact to the MHC class I molecules and of residues crucial for the recognition of peptides by the TCR. Combinatorial libraries support the design of peptides applicable in a tumor model vaccine and are also one the most promising tools in the design of therapeutic vaccines to treat incorrectly oriented immune responses.

## 3 Peptide Selection by the Transporter Associated with Antigen Processing

The crucial role of TAP in MHC class I antigen presentation became evident from studies with mutant cell lines defective in class I surface expression. This led to the identification of MHC-encoded genes for members of the ATP-binding cassette (ABC) superfamily of proteins (Deverson et al. 1990; Spies et al. 1990; Trowsdale et al. 1990). These proteins are involved in the transport of a variety of substrates across membranes and include the multidrug resistance P-glycoprotein (MDR), the cystic fibrosis transmembrane conductance regulator (CFTR), and the adreno-leukodystrophy protein (ALDP) (HIGGINS 1992). ABC transport proteins are typically composed of four domains: two transmembrane domains made up of six to ten membrane-spanning regions, as judged by hydropathy profiles, and two domains involved in nucleotide binding and hydrolysis, containing the Walker A and B motifs. TAP is a heterodimeric complex formed by TAP1 and TAP2, containing one nucleotide binding and one transmembrane domain each. It could be demonstrated by ATP-cross-linking that the hydrophilic domains of either TAP1 or TAP2 alone can indeed bind ATP (MÜLLER et al. 1994; WANG et al. 1994). TAP could be localized to the ER membrane (KLEIJMEER et al. 1992) and the stoichiometry of the purified complex was shown to be 1:1 (MEYER et al. 1994).

The function of TAP as a peptide translocator could be proven by restoration of class I surface expression in the mutant cell lines through transfection with the *tap* genes (POWIS et al. 1991; SPIES and DEMARS 1991; ATTAYA et al. 1992) and, more directly, by ATP-dependent peptide translocation into the ER lumen using semi-permeabilized cells (ANDROLEWICZ et al. 1993; NEEFJES et al. 1993) or isolated microsomes (SHEPHERD et al. 1993; MEYER et al. 1994). These experiments rely on trapping of translocated peptides in the ER by binding to class I molecules or on N-glycosylation of peptides carrying a consensus recognition sequence (NXS/T) through enzymes on the lumenal side of the ER membrane. From its central role in the antigen processing pathway, supplying peptides to many different class I molecules, the question arose whether TAP puts a restriction on the pool of peptides available for presentation to cytotoxic T cells, possibly acting as a bottleneck in the process. The important role of TAP is also highlighted by recent findings on how viruses avoid immune recognition by inhibiting steps in antigen presentation, in particular TAP (PLOEGH 1998).

## 3.1 Peptide Libraries in the Study of TAP

The first evidence for peptide selection by TAP came from a functional polymorphism in rat (POWIS et al. 1992). While expression of the $cim^a$-allele of TAP2 (for class I modifier) leads to normal loading of the class I molecule $RT1A^a$, preferentially with peptides with positively charged COOH-terminals, loading is less efficient in the presence of the $cim^u$-allele. Also, a different spectrum of peptides can be eluted from $RT1A^a$-molecules, with aliphatic or aromatic residues at the COOH-terminal. The assumption that the differences in the TAP2 alleles result in differential transport of peptides, depending on their COOH-terminal amino acid, could subsequently be confirmed by in vitro transport assays (HEEMELS et al. 1993; MOMBURG et al. 1994; SCHUMACHER et al. 1994). A study further examining this polymorphism is also the first report of the use of a partially randomized peptide library for TAP (HEEMELS et al. 1993). ER-trapped peptides from a partly degenerate library with a site for radioiodination, N-glycosylation recognition sequence and a fixed COOH-terminal amino acid (Ile, total of 384 peptides) were compared electrophoretically with those from a library with degenerate COOH-terminals (2304 peptides). The results are in line with a more systematic report using poly-alanine sequences, again containing radioiodination and glycosylation sites as well as one partially randomized position (mixture of seven peptides each) (HEEMELS and PLOEGH 1994) and the experiments using variants of high-affinity peptides (MOMBURG et al. 1994; NEEFJES et al. 1995). Thus, while *cim*$^u$ is relatively promiscuous towards the peptide COOH-terminal but is restricted to peptides 8–10 amino acids in length, *cim*$^a$ selects for hydrophobic COOH-terminals but also tolerates longer peptides (8–13 amino acids). A study trying to extend the use of the partially randomized library ($n = 2304$) to the polymorphism in mouse failed to detect any marked difference between the alleles tested (SCHUMACHER et al. 1994). This is in agreement with the absence of functional polymorphism for mouse and human TAP, as shown using high-affinity peptides or poly-alanine scans (OBST et al. 1995; DANIEL et al. 1997).

The need for radioiodination and glycosylation recognition sequences when using peptide libraries in conjunction with TAP-transport assays highlights the problems associated with these assays. Readout is mostly via the amount of glycosylated peptide accumulated in the ER, the result of a multi-step process composed of transport glycosylation and ATP-driven export of peptides and possibly degradation, each process with its characteristic kinetic. The apparent transport rate could thus be influenced by processes other than transport alone, which can be seen from differences in transport rate and efficiency to compete for transport of a reporter peptide (NEEFJES et al. 1995). The use of totally randomized peptide libraries therefore necessitates the development of a bimolecular assay measuring peptide affinity for TAP directly. This was achieved by exploiting the fact that TAP is arrested below room temperature in a conformation with the peptide binding region accessible, allowing for the determination of equilibrium binding affinity constants $K_D$ (VAN ENDERT et al. 1994). When used in conjunction with competition for binding of a reporter peptide with known affinity, this gives averaged affinity constants for peptide libraries directly. Totally randomized soluble peptide libraries were used to establish the length selectivity of human TAP (VAN ENDERT et al. 1994). Efficient competition for binding was found for peptides 8–16 amino acids in length, with 9–12 amino acids ideally suited and irrelevant of the reporter peptide used. This is consistent with peptides ideally 9–13 amino acids in length in an approach using glycosylation and radioiodination sequences placed at the $NH_2$- and COOH-terminal, thus avoiding interference from degradation products (KOOPMANN et al. 1996). The latter approach was used with defined sequences as well as with partially randomized libraries (randomized at position 3 to C-5). For rat TAP, a similar picture was found as before, but with some transport also of peptides 18–20 amino acids in length for *cim*$^a$.

## 3.2 Recognition Principle of TAP

From the studies with permutations of high-affinity peptides used in direct transport assays, rat *cim*$^u$ and mouse TAP had been grouped together as specific for hydrophobic COOH-terminals but otherwise rather promiscuous, while *cim*[a] and human TAP had been considered as being largely unspecific towards side chain replacements (Neefjes et al. 1995). Therefore, it seemed surprising that the average affinity of human TAP for a totally randomized nonamer library was found to be 17-fold lower than for the reporter peptide used (Uebel et al. 1997). Although this value is tenfold lower than for peptide-MHC class I interactions, it indicates marked selectivity for human TAP. This is also illustrated by the effects of single amino acid substitutions that resulted in only two- to threefold differences in transport rates with the high affinity peptides, but in up to 80-fold differences as revealed by the library concept. One possible explanation is that the high concentrations used over long periods of time in the transport studies resulted in saturation of the glycosylation machinery and thus in underestimation of differences of affinities. Furthermore, only variations of class I motifs had been used in these studies; thus, small effects in a high affinity context might have been overlooked.

Subsequent mapping of the fine specificity of human TAP, using nonapeptide sublibraries, has led to an understanding of the principles of substrate binding by TAP (Fig. 3), even in the absence of a high resolution crystal structure (Uebel et al. 1997). The most pronounced effects for side chain substitutions were found for the COOH-terminal. Here, aromatic, aliphatic, and positively charged residues were preferred, with negatively charged amino acids being strongly destabilizing. In addition, a $NH_2$-terminal region (positions 1–3) was found to be equally important. Although each of the three positions alone is not as relevant as the COOH-terminal, the combined effect exceeds that of the COOH-terminal. There is no clear prevailing pattern for side chains at these positions, but negatively charged residues were generally disfavored, while positively charged amino acids, in particular Arg, as well as most of the hydrophobic residues were preferred. Substitutions at sequence positions 4–8 however had only minor effects. As opposed to MHC class I molecules, TAP does not prefer only one anchor residue per position, but rather one or more classes of residues. The emerging binding motif for TAP can also be used for prediction of peptide affinities, as illustrated by the two extremes tested (NRYMPRIRY, $K_D = 137nM$; EPGNTWDED, $K_D > 1mM$).

Interestingly, Pro at position 2 had the strongest destabilizing effect of all substitutions. A similar effect for Pro had been observed before in the transport assays for mouse, human, and rat TAP at positions 2 or 3 (Heemels and Ploegh 1994; Neisig et al. 1995; van Endert et al. 1995). This, together with the need for free peptide $NH_2$- and COOH-terminals (Schumacher et al. 1994), led to the speculation that the peptide backbone might play an important role in binding to TAP. This would explain how high affinity, comparable to that of class I molecules albeit with broader selectivity, is achieved by TAP. Another indication for the involvement of the peptide backbone is the undetectably low affinity of retro-inverse peptides, i.e., all-D-amino acid variants of a peptide but with reverted sequence. These peptides can adopt the same relative side chain conformations but differ in the arrangement of main chain atoms. Experiments with nonamer peptides and peptide libraries containing the one D-amino acid position have allowed mapping of the backbone binding to the three $NH_2$-terminal positions and the

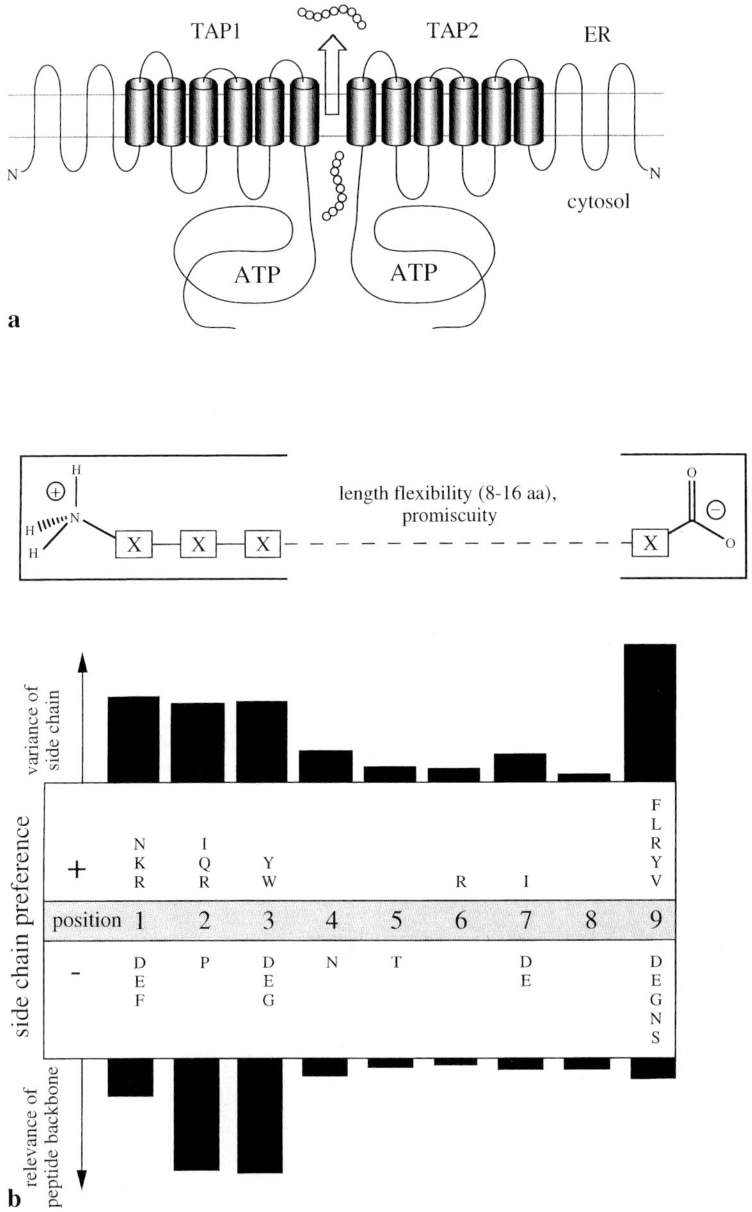


**Fig. 3.** **a** Model of the transporter associated with antigen processing (TAP) complex. **b** Peptide residues preferred (+) and disfavored (–) by human TAP are given at the individual positions of a nonamer peptide library $O/X_8$. The relevance of preferred side chains, given as the variance of the stabilizing factors and the relevance of the peptide backbone derived from D-amino acid libraries, are indicated (UEBEL et al. 1997). The peptide recognition principle of human TAP involves the three $NH_2$-terminal as well as the very COOH-terminal residue of the peptide

COOH-terminal residue (Uebel et al. 1997). In order to extend the findings on the substrate binding mechanism of TAP to longer peptides, undecapeptide libraries and 15-mer peptides with one D-position have also been tested. Again, the backbone effect maps to the $NH_2$-terminal region and the very COOH-terminal. Thus, TAP holds on to peptides via interaction with backbone and side chains at the ends, explaining the observed length selectivity. While peptides shorter than eight amino acids are too short to span the interaction sites, longer peptides can bulge out until an upper limit, possibly due to steric hindrance, is reached. This is consistent with findings that TAP can accommodate peptides with bulky, nonnatural side chains or branched peptides (Uebel et al. 1995; Grommé et al. 1997).

The COOH-terminal peptide residues preferred by TAP match those supplied by the tryptic and chymotryptic activities of the proteasome (Gaczynska et al. 1993; Ehring et al. 1996) as well as with anchors used by most human class I molecules. Aliphatic, aromatic and basic residues are preferred by both class I and TAP; for example, the COOH-terminal anchors for the most frequent human class I allele HLA-A2, Val/Leu, are among the residues strongly preferred by TAP. For the class I anchors found at positions 2 and 3, a match is less pronounced. The strongly disfavored Pro at position 2 is even found as an anchor for several HLA-B alleles. This could imply that trimming of peptides has to occur for loading onto these HLA alleles as well as onto HLA-A*0101, with the disfavored Asp/Glu as anchor at position 3, explaining why TAP can transport longer peptides than needed for direct class I loading. Recently, it was shown that the contact of a class I-bound peptide with the TCR is mainly with residues 5–8, a region where TAP exerts minimal selection (Garboczi et al. 1996; Garcia et al. 1996). It can thus be speculated that TAP has coevolved with other components of the class I pathway, displaying a mechanism for high affinity binding of peptides in order to supply peptides optimal or binding to different class I alleles without compromising T cell response.

## 4 Binding Motif of MHC Class II Molecules

Molecular interactions which determine peptide binding to MHC class II molecules are of great importance for the design of synthetic vaccines and for an improved understanding of immunologically mediated diseases. Compared to MHC class I proteins MHC class II molecules bind longer peptides (10–25 amino acids) with no apparent restriction in peptide length. Also, they show allele-specific motifs, which were not completely characterized by pool sequencing due to the difficulty in aligning peptide sequences of variable length (Rudensky et al. 1992; Chicz et al. 1993). The fine specificity of peptide binding to MHC class II molecules has been analyzed with substituted peptides (Malcherek et al. 1994), by isolation and sequencing of individual class II-associated peptides (Chicz et al. 1993), or by pool sequencing of natural ligands revealing allele-specific motifs (Falk et al. 1994). Quite different to these more classical methods is the search for MHC ligands using M13 phage display libraries (Hammer et al. 1992, 1994).

Crystal structure analysis of the human MHC class II molecule HLA-DR1 complexed with a tridecapeptide from influenza virus shows several interaction sites or pockets within the peptide binding site of HLA-DR1, five of which accommodate hydrophobic side chains of the bound influenza virus peptide (STERN et al. 1994). Many of the residues forming these pockets are highly polymorphic. This polymorphism is thought to be responsible for the different peptide specificities of different class II proteins. There are 11 core residues of the influenza virus peptide interacting with the MHC molecule.

Completely randomized synthetic peptide libraries from seven to 16 amino acids ($X_7$–$X_{16}$) were synthesized and applied to elucidate an ideal peptide length applicable in competition studies. A minimal length of 11 amino acids was determined (FLECKENSTEIN et al. 1996). Competition experiments with 220 undecapeptide amide libraries $O/X_{10}$ in total representing a combinatorial library of $2.05 \times 10^{14}$ undecapeptide amides can be considered as an attempt to define the conformational space offered by the MHC class II molecule. The analytical data from pool sequencing, mass spectrometry and amino acid analysis confirmed the close to equimolar distribution of peptides in this unique undecapeptide amide library. Tolerance to amino acid variations in the 11 sequence positions was determined by calculating the average of the absolute values of the ln(relC) for all 20 $O/X_{10}$ sublibraries of one sequence position. For sequence positions that allow close contact of the amino acids to the DR1 molecule, low tolerance with respect to biological activity to amino acid variations was expected. In contrast, amino acids variations should be better tolerated in positions projecting away from the binding cleft. The X-ray structure of the peptide in complex with DR1 shows the orientation of the central 11 residues of an influenza virus peptide. This orientation was confirmed by the results of the studies with undecapeptide libraries: sequence positions 2, 5, 7, 8, 10 and 11 show low tolerance to modifications of amino acids (Fig. 4). Positions 1, 3 and 9 are most tolerant whereas positions 4 and 6 show intermediate tolerance to exchanges of amino acids. Defined competitor undecapeptide amides were created from the activity pattern of the undecapeptide amide library (pattern for position 2 shown in the insert in Fig. 4). The most favorable amino acids for biological activity are located in those positions, which were most sensitive to amino acid exchanges (FLECKENSTEIN et al. 1996). In a second systematic study MHC class II binding peptides were also identified by random selection of amino acids from the activity pattern of the peptide library (JUNG et al. 1998). These defined peptides resulted in high competition (>90%) of a reporter peptide labeled with a fluorescence dye.

The peptide library approach proved to be successful when applied in competition studies with undecapeptide amide libraries and reporter peptides bound to MHC class II molecules. By means of competition assays using these peptide libraries the influence of every individual amino acid in all sequence positions on MHC class II binding can be easily detected and quantified. The approach will be useful for testing all MHC class II alleles and thus characterizing their binding pockets as well as binding pockets of other receptors with degenerate ligand interaction.

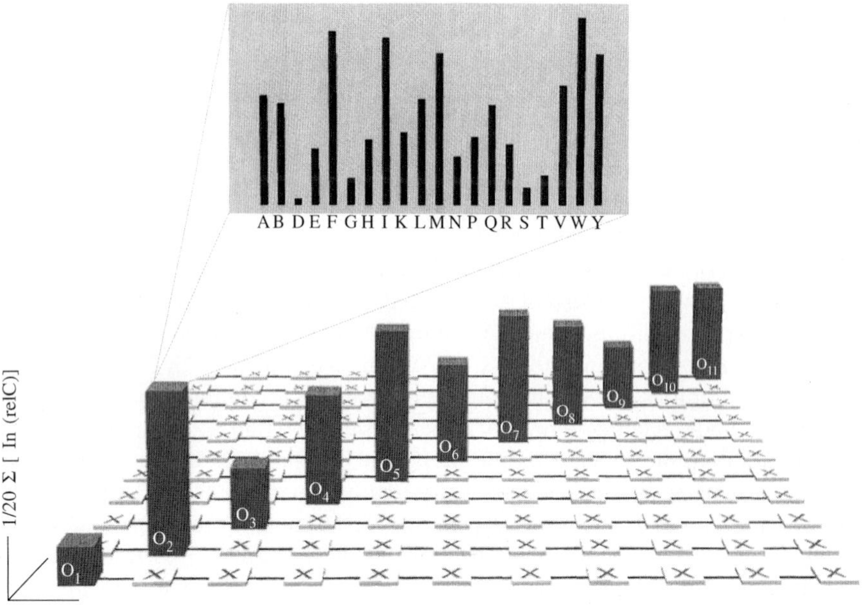


Undecapeptide amide libraries in competition experiments with MHC class II molecules

**Fig. 4.** Format of synthetic combinatorial undecapeptide amide libraries applied in competition studies with MHC class II molecules and for T helper cell proliferation studies. The library is characterized by sublibraries with one defined position O and ten randomized positions X. The level of columns indicate the tolerance to amino acid variations in the 11 sequence positions $O_{1-11}$ and represent the average of the absolute values of ln(relative competition) for the 20 sublibraries specific for one sequence position and differing in the amino acid in position O. The differences of relative competition values for the 20 amino acids O in sequence position 2 are indicated (*bars within rectangle*). The one letter code indicates L-amino acids in position O of the $XOX_9$ libraries. B (α-aminobutyric acid) was used instead of L-Cys

## 4.1 MHC Class II-Restricted T Cell Recognition

Based upon length determination using completely random $X_6$ to $X_{15}$ libraries, undecapeptide amide libraries were capable of inducing an almost maximal response on T helper cell proliferation, which was shown for the $X_{13}$ library (HEMMER et al. 1997). The highly diverse undecapeptide amide library previously used in competition studies for the identification of ligands for MHC class II molecules was thus investigated for induction of proliferation of autoreactive $CD4^+$ T cells. Cell proliferation of clone TL 5G7, which is specific for a peptide from myelin basic protein and DRB1*1501-restricted, was measured by a standard [$^3$H] thymidine-incorporation assay in the presence of peptide libraries (HEMMER et al. 1997). The results were ranked in an activity pattern (favorable amino acids listed in Table 1) of the peptide library, and individual synthetic epitopes were deduced, synthesized, and tested for proliferating activity on clone TL 5G7. All peptides induced proliferation at much lower concentration (<1ng/ml) than observed for the natural

**Table 1.** Peptide libraries in MHC class II-restricted T cell response

| Position | 1 | 2 | 3 | 4 | 5 | 6 | 7 | 8 | 9 | 10 | 11 |
|---|---|---|---|---|---|---|---|---|---|---|---|
| | D | I | I | F | F | F | K | V | V | V | V |
| | K | L | L | W | M | Y | N | N | | I | K |
| | E | V | V | M | Y | K | Y | | | | I |
| | G | M | M | V | Q | | | | | | |
| | P | F | F | L | | | | | | | |
| | | W | | I | | | | | | | |
| | | | | Q | | | | | | | |

Amino acids O in their sequence position of undecapeptide amide libraries $O/X_{10}$ favorable for proliferation of the autoreactive DRB1*1501-restricted T cell clone TL 5G7 (Hemmer et al. 1997).

epitope (100μg/ml). A protein database search for potential human or microbial antigens revealed peptides widely corresponding to the motif from the activity pattern of the library (Hemmer et al. 1997).

Homology searches provide a useful method for the identification of putative autoantigens and detecting molecular mimicry. In an independent study peptides derived from the acetylcholine receptor were found to fulfill ranking rules from the activity pattern (Jung et al. 1998). Investigation of DRB1*0301-restricted T cells derived from myasthenia graves patients with this putative autoantigen will prove a broader applicability of the approach, successful for selected clones for myelin basic protein but without any specificity in many other T cell clones investigated.

The reiterated activity of a defined amino acid residue O in neighboring positions of peptide libraries is typical for investigations with MHC class II molecules and in the proliferative response of T cell clones. We explain this observation by a translational invariance of the defined anchoring or pointing-away position. This occurs by protrusion of the terminals from the binding groove of MHC class II proteins. The ranking of amino acids in an activity pattern obtained by screening sublibraries, followed by directed or random selection of favorable amino acids for the design of new ligands, provides a comprehensive approach for the evaluation of MHC-peptide and MHC-peptide-TCR interactions. The ranking permits a prediction of peptides favorable and non-favorable for MHC binding and supporting T cell proliferation. The peptide library approach presented here offers a promising strategy to identify specific T cell activating peptides and is a promising tool to be employed in the search for immunoactive peptides and in the design of immunomodulating peptides.

## References

Androlewicz MJ, Anderson KS, Cresswell P (1993) Evidence that transporter associated with antigen processing translocate a major histocompatibility complex class I-binding peptide into the endoplasmic reticulum in an ATP-dependent manner. Proc Natl Acad Sci USA 90:9130–9134

Attaya M, Jameson S, Martinez CK, Hermel E, Aldrich C, Forman J, Fischer Lindahl K, Bavan MJ, Monaco JJ (1992) Ham-2 corrects the class I antigen-processing defect in RMA-S cells. Nature 355:647–649

Bjorkman PJ, Saper MA, Samraoui B, Bennett WS, Strominger JL, Wiley DC (1987) Structure of the human class I histocompatibility antigen HLA-A2. Nature 329:506–512

Blake J, Johnston JV, Hellström KE, Marquardt H, Chen L (1996) Use of combinatorial libraries to construct functional mimics of tumor epitopes recognized by MHC class I-restricted cytotoxic T lymphocytes. J Exp Med 184:121–130

Brown JH, Jardetzky T, Gorga JC, Stern LJ, Strominger JL, Wiley DC (1993) The three-dimensional structure of the human class II histocompatibility antigen HLA-DR1. Nature 364:33–39

Chen W, McCluskey J, Rodda S, Carbone FR (1993) Changes at peptide residues buried in the major histocompatibility complex (MHC) influence T cell recognition: a possible role for indirect conformational alterations in the MHC class I or bound peptide in determining T cell recognition. J Exp Med 177:869–873

Chicz RM, Urban RG, Gorga JC, Vignali DA, Lane WS, Strominger JL (1993) Specificity and promiscuity among naturally processed peptides bound to HLA-DR alleles. J Exp Med 178:27–47

Cresswell, P (1992) Chemistry and functional role of the invariant chain. Curr Opin Immunol 4:87–92

Daniel S, Caillatzucman S, Hammer J, Bach JF, van Endert PM (1997) Absence of functional relevance of human transporter associated with antigen-processing polymorphism for peptide selection. J Immunol 159:2350–2357

Davenport MP, Smith KJ, Barouch D, Reid SW, Bodnar WM, Willis AC, Hunt DF, Hill AVS (1997) HLA class I binding motifs derived from random peptide libraries differ at the COOH terminus from those of eluted peptides. J Exp Med 185:367–371

Degen E, Williams BD (1991) Participation of a new 88-kDa protein in the biogenesis of murine class I histocompatibility molecules. J Cell Biol 112:1099–1115

Deverson EV, Gow IR, Coadwell WJ, Monaco JJ, Butcher GW, Howard JC (1990) MHC class II region encoding proteins related to the multidrug resistance family of transmembrane transporters. Nature 348:738–741

Dooley CT, Chung NN, Wilkes BC, Schiller PW, Bidlack JM, Pasternack GW, Houghten RA (1994) An all D-amino acid opioid peptide with central analgesic activity from a combinatorial library. Science 266:2019–2022

Ehring B, Meyer TH, Eckerskorn C, Lottspeich F, Tampé R (1996) Effects of MHC-encoded subunits on the peptidase and proteolytic activities of human 20S proteasomes – Cleavage of proteins and antigenic peptides. Eur J Biochem 235:404–415

Falk K, Rötzschke O, Stevanović S, Jung G, Rammensee HG (1991) Allele-specific motifs revealed by sequencing of self peptides eluted from MHC molecules. Nature 351:290–296

Falk K, Rötzschke O, Stevanović S, Jung G, Rammensee HG (1994) Pool sequencing of natural HLA-DR, DQ and DP ligands reveals detailed peptide motifs, constraints of processing, and general rules. Immunogenetics 39:230–242

Fleckenstein B, Kalbacher H, Muller CP, Stoll D, Halder T, Jung G, Wiesmüller KH (1996) New ligands binding to the human leukocyte antigen class II molecule DRB1*0101 based on the activity pattern of an undecapeptide library. Eur J Biochem 240:71–77

Fremont DH, Matsumura M, Stura EA, Peterson PA, Wilson IA (1992) Cristal structure of two viral peptides in complex with murine MHC class I H-2K$^b$. Science 257:919–927

Furka A, Sebestyén F, Asgedom M, Dibó G (1991) General method for rapid synthesis of multicomponent peptide mixtures. Int J Pept Prot Res 37:487–493

Gaczynska M, Rock KL, Goldberg AL (1993) Gamma-interferon and expression of MHC genes regulate peptide hydrolysis by proteasomes. Nature 365:264–267

Gallop M, Barrett RW, Dower WJ, Fodor SPA, Gordon EM (1994) Applications of combinatorial technologies to drug discovery. 1. Background and peptide combinatorial libraries. J Med Chem 37:1233–1251

Garboczi DN, Ghosh P, Utz U, Fan QR, Biddison WE, Wiley DC (1996) Structure of the complex between human T cell receptor, viral peptide and HLA-A2. Nature 384:134–141.

Garcia KC, Degano M, Stanfield RL, Brunmark A, Jackson MR, Peterson PA, Teyton L, Wilson IA (1996) An $\alpha/\beta$ T cell receptor structure at 2.5Å and its orientation in the TCR–MHC complex. Science 274:209–219

Geysen HM, Rodda SJ, Mason TJ (1986) A priori delineation of a peptide which mimics a discontinuous antigenic determinant. Mol Immunol 23:709–715

Gosselin EJ, Wardwell K, Gosselin DR, Alter N, Fisher JL, Guyre PM (1992) Enhanced antigen presentation using human Fc-$\gamma$ receptor-specific immunogens. J Immunol 149:3477–3481

Grommé M, van der Valk R, Sliedregt K, Vernie L, Liskamp R, Hämmerling GJ, Koopmann JO, Momburg F, Neefjes JJ (1997) The rational design of TAP inhibitors using peptide substrate modifications and peptidomimetics. Eur J Immunol 27:898–904
Gundlach B, Wiesmüller KH, Junt T, Kienle S, Jung G, Walden P (1996a) Determination of T cell epitopes with random peptide libraries. J Immunol Meth 192:149–155
Gundlach B, Wiesmüller KH, Junt T, Kienle S, Jung G, Walden P (1996b) Specificity and degeneracy of minor histocompatibility antigen-specific MHC-restricted cytotoxic T lymphocytes. J Immunol 156:3645–3651
Hammer J, Belunis C, Bolin D, Papadopoulos J, Walsky R, Higelin J, Danho W, Sinigaglia F, Nagy, ZA (1994) High-affinity binding of short peptides to major histocompatibility complex class II molecules by anchor combinations. Proc Natl Acad Sci USA 91:4456–4460
Hammer J, Takacs B, Sinigaglia F (1992) Identification of a motif for HLA-DR1 binding peptides using M13 phage display libraries J Exp Med 176:1007–1013
Hemmer B, Fleckenstein BT, Vergelli M, Jung G, McFarland H, Martin R, Wiesmüller KH (1997) Identification of high potency microbial and self ligands for a human autoreactive class II-restricted T cell clone. J Exp Med 185:1651–1659
Heemels MT, Ploegh HL (1994) Substrate-specificity of allelic variants of the TAP peptide transporter. Immunity 1:775–784
Heemels MT, Schumacher TNM, Wonigeit K, Ploegh HL (1993) Peptide translocation by variants of the transporter associated with antigen processing. Science 262:2059–2063
Higgins CF (1992) ABC transporters: from microorganisms to man. Annul. Rev Cell Biol 8:67–113
Hill A, Ploegh H (1995) Getting the inside-out – the transporter associated with antigen-processing (TAP) and the presentation of viral-antigen. Proc Natl Acad Sci USA 92:341–343
Houghten RA, Pinilla C, Blondelle SE, Appel JR, Dooley CT, Cuervo JH (1991) Generation and use of synthetic peptide combinatorial libraries for basic research and drug discovery. Nature 364:84–86
Jung C, Kalbus M, Fleckenstein B, Melms A, Jung G, Wiesmüller KH (1998) New ligands for HLA DRB1*0301 by random selection of favourable amino acids ranked by competition studies with undecapeptide amide sublibraries. J Immunol Meth 219:139–149
Jung G (1996) Natural peptide libraries of microbial and mammalian origin. In: Jung G. (ed.) Combinatorial peptide and nonpeptide libraries – A handbook for the search of lead structures. Verlag Chemie, Weinheim, FRG
Kappler JW, Roehm N, Marrack P (1987) T cell tolerance by clonal elimination in the thymus. Cell 49:273–280
Kleijmeer M, Kelly A, Geuze HJ, Slot JW, Townsend A, Trowsdale J (1992) Location of MHC-encoded transporters in the endoplasmic reticulum and *cis*-golgi. Nature 357:342–344
Koopmann JO, Hämmerling GJ, Momburg F (1997) Generation, intracellular-transport and loading of peptides associated with MHC class-I molecules. Curr Opin Immunol 9:80–88
Koopmann JO, Post M, Neefjes JJ, Hämmerling GJ, Momburg F (1996) Translocation of long peptides by transporters associated with antigen-processing (TAP). Eur J Immunol 26:1720–1728
Lam KS, Salmon SE, Hersh EM, Hruby VJ, Kazmierski WM, Knapp RJ (1991) A new type of synthetic peptide library for identifying ligand-binding activity. Nature 364:82–84
Madden DR, Garboczi DN, Wiley DC (1993) The antigenic identity of peptide-MHC complexes: a comparison of the conformations of five viral peptides presented by HLA-A2. Cell 75:693–708
Malcherek G, Gnau V, Stevanović S, Rammensee HG, Jung G, Melms A (1994) Analysis of allele-specific contact sites of natural HLA-DR17 ligands. J Immunol 153:1141–1149
Matsumura M, Saito Y, Jackson MR, Song ES, Peterson PA (1992) In vitro peptide binding to soluble empty class I major histocompatibility complex molecules isolated from transfected Drosophila melanogaster cells. J Biol Chem 267:23589–23595
Metzger JW, Kempter C, Wiesmuller KH, Jung G (1994) Electrospray mass spectrometry and tandem mass spectrometry of synthetic multicomponent peptide mixtures: determination of composition and purity. Anal Biochem 219:261–277
Meyer TH, van Endert PM, Uebel S, Ehring B, Tampé R (1994) Functional expression and purification of the ABC transporter complex-associated with antigen-processing (TAP) in insect cells. FEBS Lett 351:443–447
Momburg F, Roelse J, Howard JC, Butcher GW, Hämmerling GJ, Neefjes JJ (1994) Selectivity of MHC-encoded peptide transporters from human, mouse and rat. Nature 367:648–651
Müller KM, Ebensperger C, Tampé R (1994) Nucleotide binding to the hydrophilic COOH-terminal domain of the transporter associated with antigen processing (TAP). J Biol Chem 269:14032–14037

Neefjes JJ, Gottfried E, Roelse J, Grommé M, Obst R, Hämmerling GJ, Momburg F (1995) Analysis of the fine specificity of rat, mouse and human TAP peptide transporters. Eur J Immunol 25:1113–1136
Neefjes JJ, Momburg F, Hämmerling GJ (1993) Selective and ATP-dependent translocation of peptides by the MHC-encoded transporter. Science 261:769–771
Neisig A, Roelse J, Sijts AJA, Ossendorp F, Feltkamp MCW, Kast WM, Melief CJM, Neefjes JJ (1995) Major differences in transporter associated with antigen presentation (TAP)-dependent translocation of MHC class I-presentable peptides and the effect of flanking sequences. J Immunol 154:1273–1279
Obst R, Armandola EA, Nijenhuis M, Momburg F, Hämmerling GJ (1995) TAP polymorphism does not influence transport of peptide variants in mice and humans. Eur J Immunol 25:2170–2176
Ortmann B, Copeman J, Lehner PJ, Sadasivan B, Herberg JA, Grandea Ag, Riddell SR, Tampé R, Spies T, Trowsdale J, Cresswell P (1997) Science 277:1306–1309
Pieters J, Horstmann H, Bakke O, Griffiths G, Lipp J (1991) Intracellular transport and localization of major histocompatibility complex class II molecules and associated invariant chain. J Cell Biol 115:1213–1223
Pinilla C, Appel JR, Blanc P, Houghten RA (1992) Rapid identification of high-affinity peptide ligands using positional scanning synthetic peptide combinatorial libraries. BioTechniques 13:901–905
Ploegh HL (1998) Viral strategies of immune evasion. Science 280:248–253
Powis SJ, Deverson EV, Coadwell WJ, Ciruela A, Huskisson NS, Smith H, Butcher GW, Howard JC (1992) Effect of polymorphism of an MHC-linked transporter on the peptides assembled in a class I molecule. Nature 357:211–215
Powis SJ, Townsend ARM, Deverson EV, Bastin J, Butcher GW, Howard JC (1991) Restoration of antigen presentation to the mutant cell line RMA-S by an MHC-linked transporter. Nature 354:528–531
Pridzun L, Wiesmüller KH, Kienle S, Jung G, Walden P (1996) Amino acid preferences in the otapeptide subunit of the major histocompatibility complex class I heterotrimer H-2L$^{d}$. Eur J Biochem 236:249–253
Rammensee HG, Falk K, Rötzschke O (1993) MHC molecules as peptide receptors. Curr Opin Immunol 5:35–44
Rammensee HG, Friede T, Stevanović S (1995) MHC ligands and peptide motifs: First listing. Immunogenetics 41:178–222
Rock KL, Gramm C, Rothstein L, Clark K, Stein R, Dick L, Hwang D, Goldberg AL (1994) Inhibitors of the proteasome block the degradation of most cell-proteins and the generation of peptides presented on MHC class-I molecules. Cell 78:761–771
Rudensky AY, Preston-Hurlburt P, Al-Ramadi BK, Rothbard J, Janeway CA (1992) Truncation variants of peptides isolated from MHC class II molecules suggest sequence motifs. Nature 359:429–431
Sadasivan B, Lehner PJ, Ortmann B, Spies T, Cresswell P (1996) Roles for calreticulin and a novel glycoprotein, tapasin, in the interaction of MHC class-I molecules with TAP. Immunity 5:103–114
Sant A, Miller J (1994) MHC class II antigen processing: biology of invariant chain. Curr Opin Immunol 6:57–63
Schumacher TN, Kantesaria DV, Heemels MT, Ashton-Rickardt PG, Shepherd JC, Früh K, Yang Y, Peterson PA, Tonegawa S, Ploegh HL (1994) Peptide length and sequence specificity of the mouse TAP1/TAP2 translocator. J Exp Med 179:533–540
Schumacher TNM, Kantesaria DV, Serreze DV, Roopenian DC, Ploegh HL (1994) Transporters from H-2(b), H-2(d), H-2(s), H-2(k), and H-2(g7) (NOD/lt) haplotype translocate similar sets of peptides. Proc Natl Acad Sci USA 91:13004–13008
Schwartz RH (1985) T lymphocyte recognition of antigen in association with gene products of the major histocompatibility complex. Ann Rev Immunol 3:261–327
Shepherd JC, Schumacher TN, Ashton-Rickardt PG, Imaeda S, Ploegh HL, Janeway CAJ, Tonegawa S (1993) TAP1-dependent peptide translocation in vitro is ATP dependent and peptide selective. Cell 74:577–584
Sherman MA, Weber DA, Jensen PE (1995) DM enhances peptide binding to class II MHC by release of invariant chain-derived peptide. Immunity 3:197–205
Singer A, Hodes RJ (1983) Mechanism of T cell B cell interaction. Ann Rev Immunol 1:211–241
Stevanović S, Jung G (1993) Multiple sequence analysis: Pool sequencing of synthetic and natural peptide libraries. Anal Biochem 212:212–220
Stern LJ, Brown JH, Jardetzky TS, Gorga JC, Urban RG, Strominger JL, Wiley DC (1994) Crystal structure of the human class II MHC protein HLA-DR1 complexed with an influenza virus peptide. Nature 368:215–221
Stevens J, Wiesmüller KH, Barker PJ, Walden P, Butcher G, Joly E (1998a) Efficient generation of MHC class I-peptide complexes using synthetic peptide libraries. J Biol Chem 274:2874–2884

Spies T, Bresnahan M, Bahram S, Arnold D, Blanck G, Mellins E, Pious D, DeMars R (1990) A gene in the human major histocompatibility complex class II region controlling the class I antigen presentation pathway. Nature 348:744–47

Spies T, DeMars R (1991) Restored expression of major histocompatibility class I molecules by gene transfer of a putative peptide transporter. Nature 351:323–324

Sugita M, Brenner MB (1994) An unstable beta-2-microglobulin – major histocompatibility complex class-I heavy-chain intermediate dissociates from calnexin and then is stabilized by binding peptide. J Exp Med 180:2163–2171

Townsend A, Bodmer H (1989) Antigen recognition by class I-resticed T lymphocytes. Ann Rev Immunol 7:601–624

Trowsdale J, Hanson I, Mockridge I, Beck S, Townsend A, Kelly A (1990) Sequences encoded in the class II region of the MHC related to the 'ABC' superfamily of transporters. Nature 348:741–744

Udaka K, Wiesmüller KH, Kienle S, Jung G, Walden P (1995a) Tolerance to amino acid variations in peptides binding to the MHC class I protein H-2K$^b$. J Biol Chem 720:24130–24136

Udaka K, Wiesmüller KH, Kienle S, Jung G, Walden P (1995b) Decrypting the structure of MHC-I restricted CTL epitopes with complex peptide libraries. J Exp Med 181:2097–2108

Udaka K, Wiesmüller KH, Kienle S, Jung G, Walden P (1996) Self MHC-restricted peptides recognized by an alloreactive T lymphocyte clone. J Immunol 157:670–678

Uebel S, Kraas W, Kienle S, Wiesmüller KH, Jung G, Tampé R (1997) Recognition principle of the TAP-transporter disclosed by combinatorial peptide libraries. Proc Natl Acad Sci USA 94:8976–8981

Uebel S, Meyer TH, Kraas W, Kienle S, Jung G, Wiesmüller KH, Tampé R (1995) Requirements for peptide binding to the human transporter associated with antigen-processing revealed by peptide scans and complex peptide libraries. J Biol Chem 270:18512–18516

van Endert PM, Riganelli D, Greco G, Fleischhauer K, Sidney J, Sette A, Bach JF (1995) The peptide-binding motif for the human transporter associated with antigen-processing. J Exp Med 182:1883–1895

van Endert PM, Tampé R, Meyer TH, Tisch R, Bach JF, McDevitt HO (1994) A sequential model for peptide binding and transport by the transporters associated with antigen processing. Immunity 1:491–500

Vidard L, Rock KL, Benacerraf B (1992) Diversity in MHC class II ovalbumin epitopes generated by distinct proteases. J Immunol 149:498–504

Wang KN, Früh K, Peterson PA, Yang Y (1994) Nucleotide-binding of the COOH-terminal domains of the major histocompatibility complex-encoded transporter expressed in Drosophila melanogaster cells. FEBS Lett 350:337–341

Wiesmüller KH, Feiertag S, Fleckenstein B, Kienle S, Stoll D, Herrmann M, Jung G (1996) Peptide and Cyclopeptide Libraries: Automated Synthesis, Analysis and Receptor Binding Assays. In: Jung G (ed) Combinatorial peptide and nonpeptide libraries – a handbook for the search of lead structures. Verlag Chemie, Weinheim, Germany 1996

Yewdell JW, Bennink JR (1992) Cell biology of antigen-processing and presentation to major histocompatibility complex class-I molecule-restricted T lymphocytes. Adv Immunol 52:1–123

# Antigen Sequence- and Library-Based Mapping of Linear and Discontinuous Protein–Protein-Interaction Sites by Spot Synthesis

U. Reineke[2], A. Kramer[1], and J. Schneider-Mergener[1]



## 1 Introduction

Peptides synthesized on continuous cellulose membranes by the spot synthesis technique (Frank et al. 1992) have been increasingly used to study molecular recognition events. The application of these positionally addressable peptide libraries include investigating protein/protein (literature cited throughout the text), protein/DNA (Kramer et al. 1993; Reuter et al. 1998) and protein/metal interactions (Malin et al. 1995). In addition, substrate specificities of kinases (Toomik et al. 1996; Tegge et al. 1998; Mukhija et al. 1998), proteases (Duan and Laursen 1994; Kramer et al. 1998; Reineke et al. 1999) and chaperones (Rüdiger et al. 1997) have been determined. In this chapter we review protein sequence- and library-based approaches to map linear and discontinuous antibody/antigen and receptor/ligand contact sites using the spot synthesis technique.

[1] Institut für Medizinische Immunologie, Universitätsklinikum Charité, Humboldt-Universität zu Berlin, D-10117 Berlin, Germany
[2] *Present address*: Jerini Bio Tools GmbH, Rudower Chaussee 29, D-12489 Berlin, Germany

## 2 Spot Synthesis

Spot synthesis is an easy and flexible technique for simultaneous, parallel chemical synthesis on membrane supports (for detailed experimental description see Frank 1992; Kramer et al. 1994; Frank and Overwin 1996; Kramer and Schneider-Mergener 1998). The most frequent application is the synthesis of peptides on cellulose membranes (Fig. 1) which initially involves derivatizing the hydroxyl functions of a commercially available filter paper (e.g., Whatman, Maidstone, UK) with 9-fluorenylmethoxycarbonyl-β-alanine (Fmoc-β-Ala) and the subsequent removal of the Fmoc-group. Synthesis areas (spots) are defined by spotting a Fmoc-β-alanine-pentafluorophenyl ester solution to distinct sites on the membrane. This is done either by a pipetting robot (Abimed GmbH, Langenfeld, Germany) or manually. Blocking (acetylating) the remaining amino functions between the spots provides up to 8000 discrete reaction sites on a 20×30cm membrane for further standard, solid-phase peptide synthesis using amino acid pentafluorophenyl esters. For the synthesis of randomized positions in combinatorial libraries equimolar mixtures of amino acids are applied (Kramer et al. 1994; Kramer and Schneider-Mergener 1998). Routinely, peptides up to a length of 20 residues, but also longer peptides (Molina et al. 1996), can be synthesized with sufficient fidelity. The peptides can then be used for binding and enzymatic assays directly on the cellulose membranes or cleaved from the solid support by the use of ammonia or via special linker molecules between the membrane and the peptide (Hoffmann and Frank 1994; Volkmer-Engert et al. 1997).

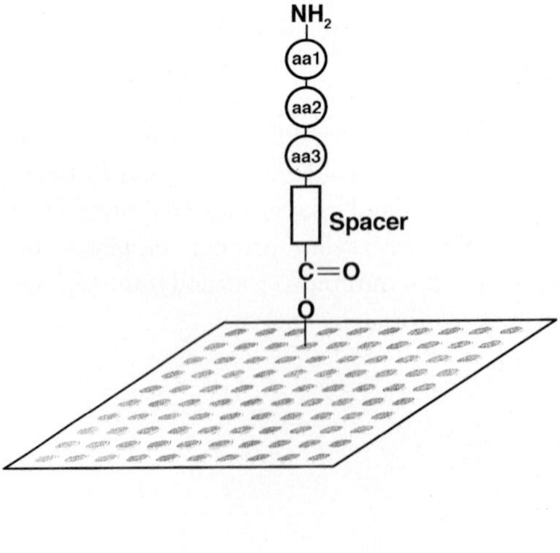


**Fig. 1.** Spot synthesis. The peptides are COOH-terminally covalently linked to the hydroxyl functions on predefined areas (*spots*) of the cellulose membrane. The arrangement of peptides on the membrane is created by the program SpotLab (Wu and Schneider-Mergener, unpublished). The synthesis is performed semi-automatically (Abimed GmbH, Langenfeld, Germany). For sufficient flexibility during synthesis and for binding studies, a two β-alanine residue spacer is inserted between the cellulose and the peptide

## 3 Overlapping Peptide Scans for the Mapping of Linear Epitopes

The site on a protein surface that is recognized by its binding partner can be a linear stretch of amino acids or a more complex discontinuous (non-linear) epitope composed of at least two segments close in space but separated on the primary sequence (Fig. 2). In linear epitopes the residues which are in contact with the binding partner comprise up to 15, but usually 8–12, amino acids. Peptides covering such epitopes are able to compete with the protein-protein interaction and usually have affinities in the same range as the native protein.

The mapping of linear epitopes using overlapping peptides derived from the protein sequence is nowadays an easy and straight forward approach (GEYSEN et al. 1984; HOUGHTEN 1985; FRANK 1992). The entire sequence is scanned with short overlapping peptides which are subsequently probed for binding to the respective protein. The sequence common to the interacting peptides is the linear epitope (Fig. 2). In this case the length of the scanning peptides should be at least 12 amino acids. In most linear epitopes not all residues are in contact with the binding partner. In order to identify side chains which are important for the interaction substitutional analysis membranes are synthesized. Each position of the epitope is replaced individually by all other amino acids and the binding properties of the single substitution analogs are tested (Fig. 3). The side chains of residues which cannot be substituted, or at least only by physico-chemically similar amino acids, are considered to be key residues in the peptide/protein and thus in the native protein/protein complex.

Sometimes characteristic patterns of key residues are observed, from which the secondary structure of the epitope in the context the entire protein can be deduced. In the right part of Fig. 3 a typical pattern for a helical epitope is shown. This epitope of human interleukin-10 (hIL-10), recognized by the monoclonal antibody CB/RS/3 (REINEKE et al. 1998a), has key residues at a distance of three to four amino acids and the presence of proline leads to complete loss of binding, even at positions not sensitive to other substitutions. In this case the key residues are

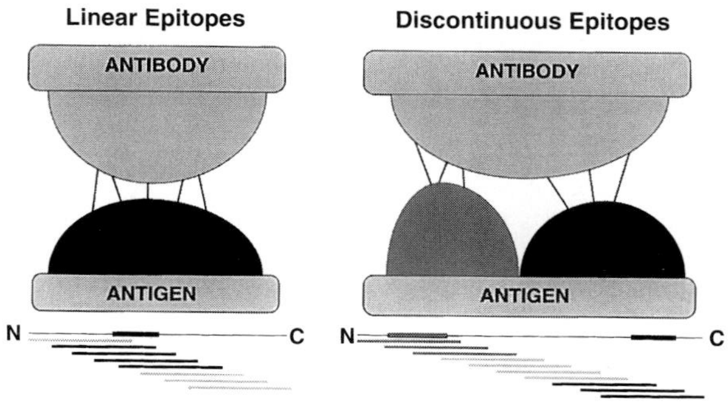


**Fig. 2.** Scheme of mapping linear and discontinuous antibody epitopes

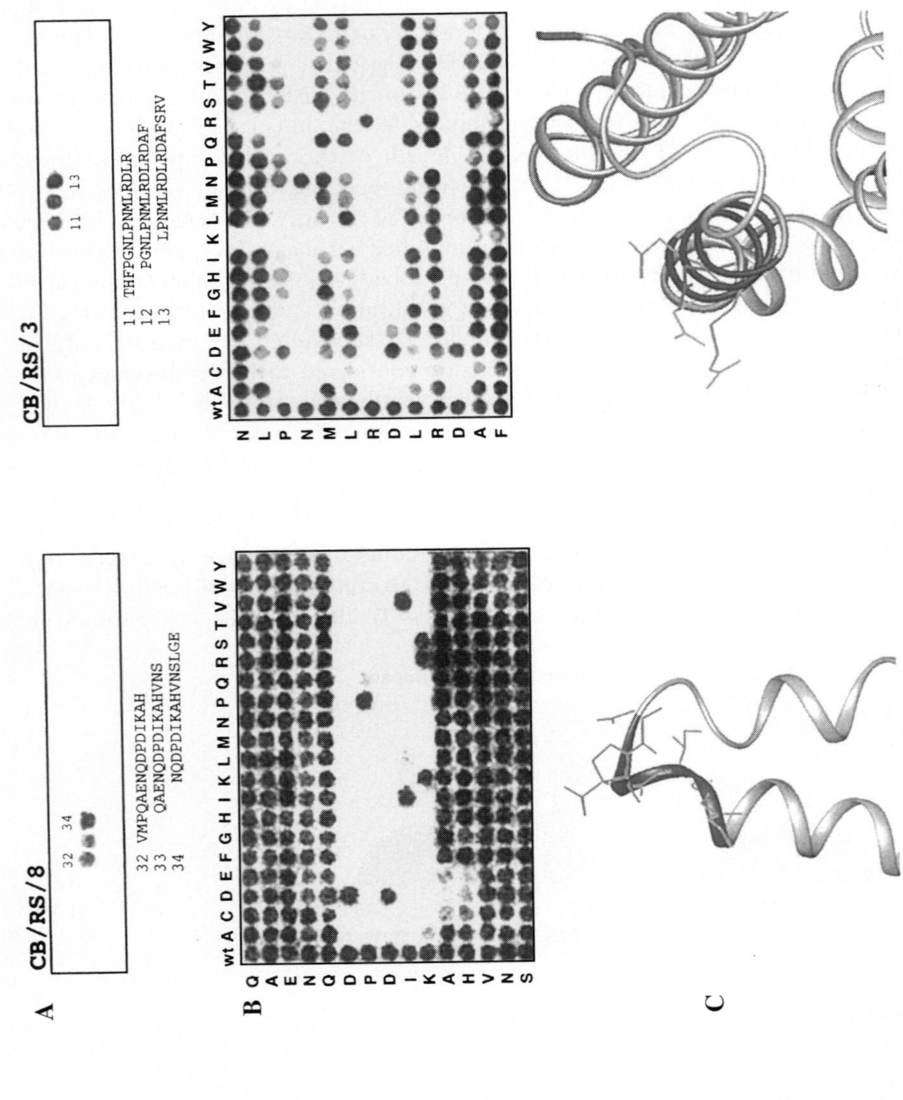
A
CB/RS/8
32 34
32 VMPQAENQDPDIKAH
33 QAENQDPDIKAHVNS
34 NQDPDIKAHVNSLGE
CB/RS/3
11 13
11 THFPGNLPNMLRDLR
12 PGNLPNMLRDLRDAF
13 LPNMLRDLRDAFSRV
B
wt A C D E F G H I K L M N P Q R S T V W Y
Q A E N Q D P D I K A H V N S
wt A C D E F G H I K L M N P Q R S T V W Y
N L P N M L R D L R D A F
C

located on the solvent exposed side of a helix (Fig. 3C). On the left panel of Fig. 3A typical substitutional analysis pattern for a loop region recognized by the anti-hIL-10 antibody CB/RS/8 is shown. Five subsequent key residues are observed, one of them being proline which stabilizes the loop in the three-dimensional hIL-10 structure.

# 4 Mapping of Discontinuous Epitopes

Since protein/protein interactions are usually mediated by relatively large contact sites (JONES and THORNTON 1996), in the majority of cases proteins recognize discontinuous sites on their binding partners (Fig. 2). The mapping of discontinuous epitopes using fragments of proteins generated either chemically or biologically (MORRIS 1996) suffers from the drawback that peptides derived from single binding regions of a discontinuous binding site generally have very low binding partner affinities which are usually not measurable in ELISA or BIAcore experiments. Therefore two problems have to be addressed in order to identify such epitopes: (1) The binding assay has to be sufficiently sensitive and (2) unspecific peptide/protein interactions which are inherent in low affinity ranges have to be avoided or ruled out. Within the last few years, several examples of investigations into discontinuous protein/protein interactions using peptides synthesized on continuous surfaces have been described. In terms of sensitivity the spot approach is especially suitable since the high peptide density on the membranes (50nmol/cm$^2$) simulates a high local concentration.

## 4.1 Sensitivity of the Assay System

Binding of a protein to cellulose-bound peptides can be detected directly on the membrane using a specific antibody which is enzyme-labeled or recognized by a second enzyme-labeled antibody. Low affinity peptide/protein interactions are therefore difficult to identify since the binding equilibrium of the peptide/protein complex on the membrane is shifted towards the uncomplexed molecules during incubation with the detection antibodies. This can be avoided by electrotransfer of peptide-bound protein onto polyvinylene difluoride membranes prior to immuno-

**Fig. 3A–C.** Mapping of the two interleukin-10 epitopes. **A** Binding of the anti-hIL-10 antibodies CB/RS/8 and CB/RS/3 to hIL-10-derived overlapping peptides (15-mers shifted by three amino acids, 146 peptides in total, starting at the *upper left side* of the membrane, 20 spots per row). Binding experiments were performed and detected using a peroxidase-conjugated anti-mouse antibody as previously described (REINEKE et al. 1998a). **B** Substitutional analyses of the epitope peptides. All residues of the wild type epitope (*left column*) are substituted by all other L-amino acids (*rows*) and tested for binding. **C** Parts of the hIL-10 structure (ZDANOV et al. 1996). The epitopes are highlighted in *black* and key residues are displayed with their side chains

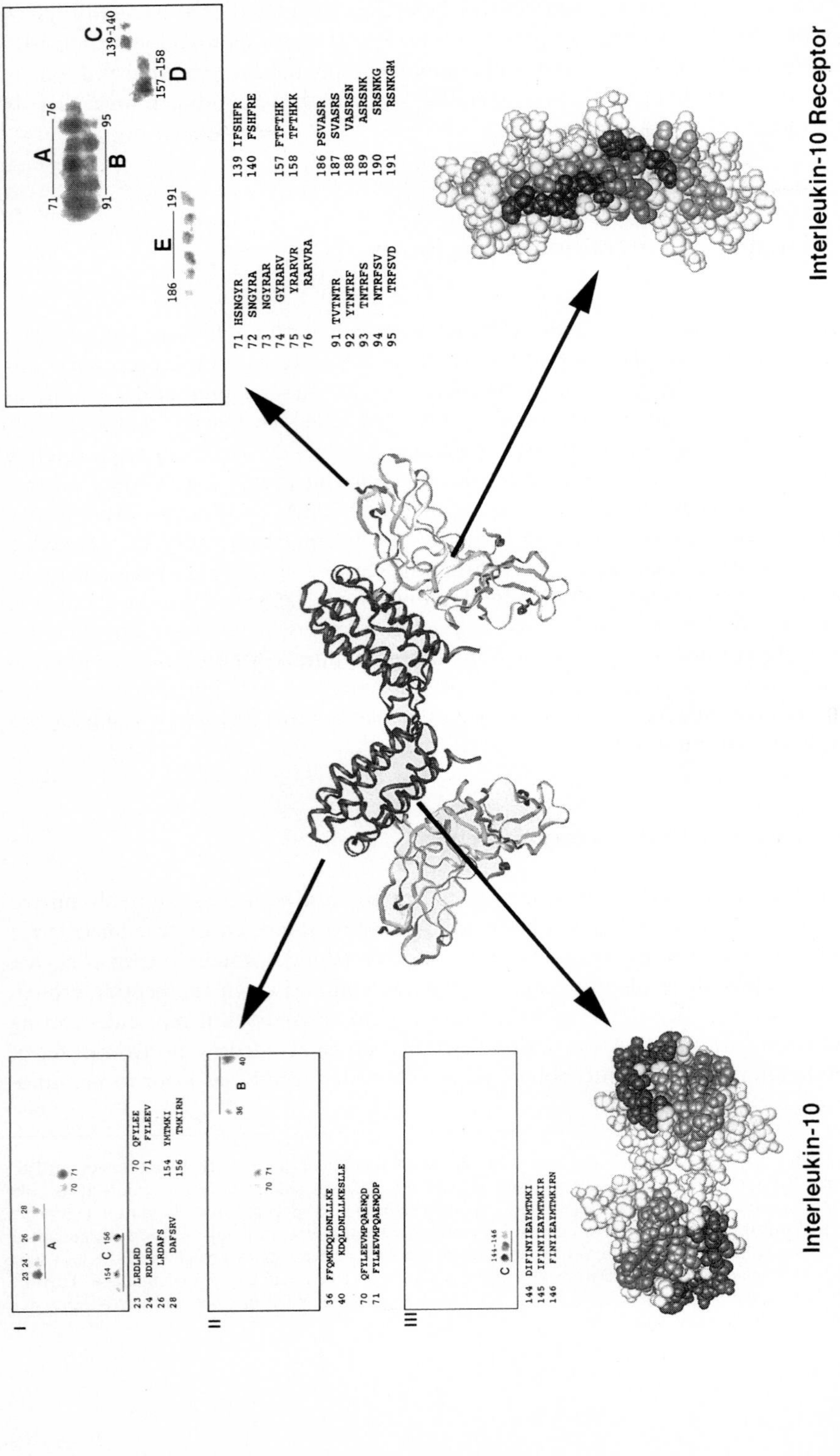
A 71 76
B 91 95
C 139-140
D 157-158
E 186 191
71 HSNGYR
72 SNGYRA
73 NGYRAR
74 GYRARV
75 YRARVR
76 RARVRA
91 TVTNTR
92 YTNTRF
93 TNTRFS
94 NTRFSV
95 TRFSVD
139 IFSHFR
140 FSHFRE
157 FTFTHK
158 TFTHKK
186 PSVASR
187 SVASRS
188 VASRSN
189 ASRSNK
190 SRSNKG
191 RSNKGM
Interleukin-10 Receptor
I
23 LRDLRD
24 RDLRDA
26 LRDAFS
28 DAFSRV
70 QFYLEE
71 FYLEEV
154 YMTMKI
156 TMKIRN
II
36 FFQMKDQLDNLLLKE
40 KDQLDNLLLKESLLE
70 QFYLEEVMPQAENQD
71 FYLEEVMPQAENQDP
III
144 DIFINYIEAYMTMKI
145 IFINYIEAYMTMKIR
146 FINYIEAYMTMKIRN
Interleukin-10

detection (Rüdiger et al. 1997; Reineke et al. 1998a). Such a procedure can be carried out in a fractionated manner which sometimes results in different binding patterns presumably due to different affinities. In this way the interaction of hIL-10 with its receptor (hIL-10R) (Fig. 4) as well as the discontinuous epitopes recognized by an anti-hIL-10 and an anti-hIL-10R antibody were mapped (Reineke et al. 1998a). In this study hIL-10 and hIL-10R scans with 6-mer and 15-mer peptides were used for the mapping the hIL-10/hIL-10R combining site. Some binding regions (region C on hIL-10, Fig. 4) were detected in both scans, whereas others (hIL-10 region B and hIL-10R regions A–D) were recognized only as 15-mer or 6-mer peptides, respectively. This is most likely due to the different lengths of binding regions. If short binding regions of a few amino acids have to be detected, the sequences which are adjacent in the primary structure but possibly buried in the protein can influence the peptide/protein interaction if synthesized as peptides and thus being accessible for the binding partner. Therefore, peptide scans with shorter peptides can help to reduce the effects of the artificial assay system. On the other hand the detection of extended binding regions may require scans of longer peptides. Another approach to increase the assay sensitivity is the use of radioactively labeled proteins as was used for the mapping of the human interleukin-6 (hIL-6)/human interleukin-6 receptor (hIL-6R) combining site with [$^{125}$I]hIL-6 and [$^{125}$I]hIL-6R (Weiergräber et al. 1995).

## 4.2 Identification of Unspecific Protein–Protein Interactions

Different strategies to distinguish between specific and unspecific peptide/protein interactions can be applied: (1) If the three-dimensional structure is available, the location of the binding regions may be informative since specific binding regions must be on the protein surface and form a connected epitope. Such structural aspects have been regarded using X-ray crystal structures or three-dimensional models for the mapping of the hIL-10/hIL-10R interaction (Reineke et al. 1998a), the hIL-6/hIL-6R combining site (Weiergräber et al. 1995), the epitope of an antibody specific for the pathological form of the bovine prion protein (Korth et al. 1997), the paratope of the anti-lysozyme antibody HyHEL-5 and an anti-angiotensin II antibody (Laune et al. 1997), the epitopes of the polyspecific antibody CB/

◄

**Fig. 4.** Mapping of the hIL-10/hIL-10R combining site. *Center*, Model of the complex between the hIL-10 homodimer (*black*) and two identical hIL-10 receptors (*gray*) (Zdanov et al. 1996). *Upper left*, Binding of the extracellular soluble part of hIL-10R (shIL-10R) to IL-10-derived peptide scans (I, 6-mers; II and III 15-mers). The IL-10 sequence was scanned with peptides (shifted by one amino acid starting at the *upper left side* of the membrane). Binding was detected after electrotransfer of cellulose-bound sIL-10R to a PDVF membrane (Rüdiger et al. 1997; Reineke et al. 1998a). *I* represents the first blot of the 6-mer peptide scan, *II* the first and *III* the third blot of the 15-mer peptide scan. Binding regions *A*, *B* and *C* of the IL-10R binding site are marked on the hIL-10 structure (*lower left*). Binding to peptides 70 and 71 is unspecific (Reineke et al. 1998a). *Upper right*, Binding of hIL-10 to a shIL-10R-derived peptide scan. ShIL-10R was scanned with 6-mer peptides shifted by one amino acid. Binding regions A–E of the IL-10 binding site are marked on the hIL-10R model

RS/5 against hIL-10 and tumor necrosis factor α (TNF-α) (Reineke et al. 1996), the TNF-α binding site on the 55-kDa TNF receptor (Reineke et al. 1996) and the epitope of the anti-lysozyme antibody D1.3 (Reineke et al. 1998b). Comparing these results with the X-ray structures of the scanned proteins alone or in a complex with its binding partner has clearly demonstrated the usefulness this technique for the mapping of discontinuous epitopes. (2) Unspecific peptide/protein interactions can furthermore be ruled out by measuring the inhibitory constant of the corresponding soluble peptides on the protein–protein interaction. If the inhibitory constant is in the same range as the dissociation constant of the peptide/protein complex, binding to the paratope is confirmed. This strategy, which can only be applied if the affinity is measurable in ELISA or Biacore studies, helped to distinguish between hIL-10 and TNF-α derived peptides binding to the paratope (specific) or to the constant region (unspecific) of monoclonal antibody CB/RS/5 (Reineke et al. 1996). (3) Similarly, specific vs unspecific results may be distinguished by incubating of the peptide membrane with related proteins. For antibody epitope mapping the antigen-derived peptide scans should be incubated with other antibodies of the same subclass but different specificities (Laune et al. 1997; Reineke et al. 1998a; Gao and Esnouf 1996). (4) Unspecific interactions can also derive from sequence similarities within the protein. A sequence which is responsible for the specific interaction with the binding partner may be partly homologous to another site within the same protein, where it is most likely not recognized in the complete protein, but may be in the peptide scan. Sequence or key residue pattern comparisons of the individual binding regions are therefore helpful in the mapping. (5) In many cases hydrophobic interactions with residues from the protein core, which only become accessible in the peptide scan, are the reason for unspecific peptide/protein binding. In contrast to other interactions such as hydrogen bonds, electrostatic and van der Waals interactions, these hydrophobic interactions can be reduced by decreasing the temperature during incubation. For instance, unspecific interactions described in the mapping of the lysozyme/mabD1.3 interaction (Reineke et al. 1998b) at room temperature were completely eliminated after incubation at 4°C.

## 5 Combinatorial Peptide Libraries for the Mapping of Linear Epitopes

If the binding partner of an antibody or a receptor is unknown, one possibility of identifying a potential binding partner is the use of biologically or chemically generated peptide libraries. This facilitates the screening of huge repertoires ($10^6$–$10^8$) of peptides for potential binding partners. Synthetic combinatorial peptide libraries were described for the first time in 1986 by Geysen et al. Since then different methods have been developed for the preparation and screening of these libraries (for reviews see Pavia et al. 1993; Strop 1994; Cortese 1996; Jung 1996; Cabilly 1998). Here we describe the synthesis of combinatorial libraries on con-

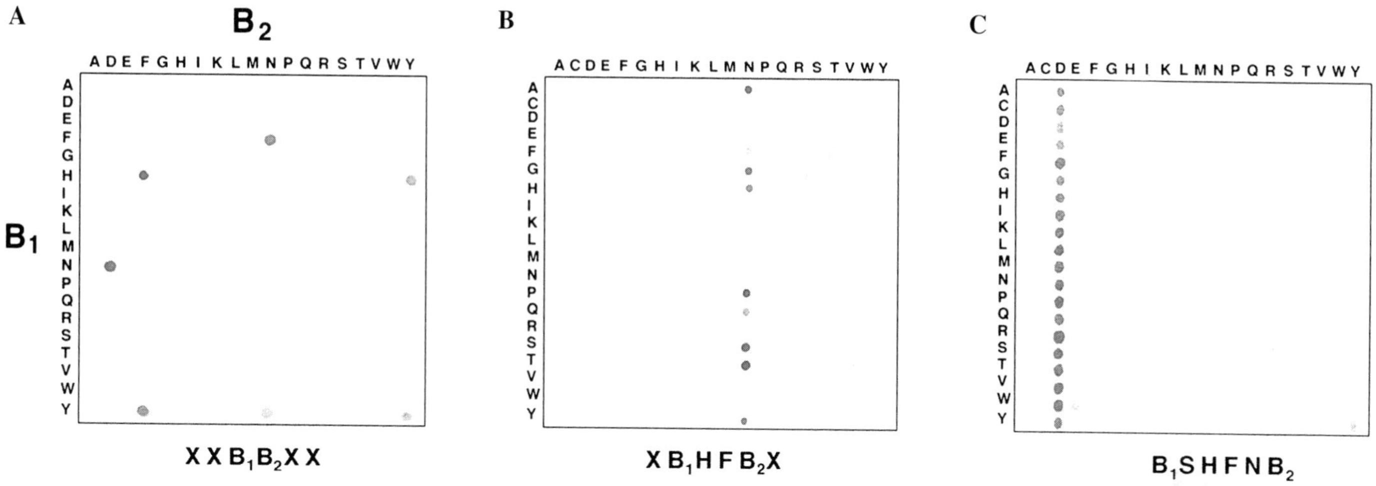


**Fig. 5A–C.** Identification of a transforming growth factor (TGF)-α epitope with a combinatorial hexapeptide library $XXB_1B_2XX$. Three steps are needed for the iterative identification of single hexapeptides binding to the anti- TGF-α antibody Tab2. This recognizes the epitope VSHFND corresponding to the $NH_2$-terminal of TGF-α. **A** Binding of Tab2 to the library $XXB_1B_2XX$ (400 ($20^2$) peptide mixtures (spots) in total). Each spot theoretically consists of 160,000 ($20^4$) peptides. In this case binding was detected with an alkaline phosphatase-conjugated anti-mouse antibody (Kramer et al. 1993). **B** Second iterative screening step. After quantification of binding, the strongest binding spot XXHFXX was selected for defining two more X-positions synthesizing the library $XB_1HFB_2X$. **C** Final screening step. After selection of the spot XSHFNX the library $B_1SHFNB_2$ now consisting of 400 single peptides was prepared and tested for binding tinuous cellulose membrane supports for the mapping of antibody epitopes (Kramer and Schneider-Mergener 1998). The general principle of combinatorial peptide libraries is the following: they are collections of peptide mixtures containing randomized positions (X) as well as defined positions (B). A combinatorial hexapeptide library $XXB_1B_2XX$, for example, consists of four randomized positions where all 20 amino acids are present, although cysteine is usually excluded, and two defined positions $B_1$ and $B_2$ (all 400 ($20^2$) combinations of the 20 amino acids). These 400 mixtures are screened for biological activity. The best mixture is chosen for a subsequent synthesis of a new library, in which the identified amino acids are fixed and two other positions are defined. Screening this library, again selecting the best mixture and defining the remaining two positions in an analogous manner, leads to the identification of the most active peptide sequence (Fig. 5).

Combinatorial peptide libraries have been developed for the de novo identification of ligand binding peptides in cases where no binding partner is known. Nevertheless, for the evaluation of newly constructed peptide libraries, several reports describe the screening with antibodies recognizing previously known linear epitopes as model systems. Identification of the TGF-α epitope VSHFND binding to the antibody Tab2 (Hoeprich et al. 1989) was achieved using a cellulose-bound hexapeptide library described above (Fig. 5) (Kramer et al. 1993, 1994). This epitope could also be mapped using an amino acid cluster library in which the defined positions B contain not only one amino acid, but are clusters of physicochemically related amino acids (Kramer et al. 1995), thus reducing the synthesis effort, i.e. the number of peptide mixtures prepared. However, attempts to use the combinatorial hexapeptide library $XXB_1B_2XX$ for the identification of other antibody epitopes mostly failed due to the length of the peptides, their relatively low complexity and bad signal to noise ratio. More complex libraries had to be constructed in order to increase the probability of detecting signals in the first screening step. For example, a combinatorial library $XXXXB_1B_2B_3XXXX$ containing three defined positions B and consisting of 8,000 peptide mixtures (spots) was used to identify the epitope LQDPRVRGLY located in the preS2 region of the hepatitis B virus (Kramer and Schneider-Mergener 1995).

From both library and substitutional analyses two major points became obvious: (1) Linear epitopes are functionally discontinuous since they contain only a few really important residues for antibody binding, whereas at other positions substitutions are allowed. These key binding residues are frequently not in adjacent positions (Geysen et al. 1988; Pinilla et al.,1993; Dong et al. 1999; Dong et al. unpublished results). (2) Consequently, taking these structural requirements of antibody-epitope interactions into account, appropriate distribution of the defined positions in combinatorial peptide libraries should improve the quality of the library. Combining the combinatorial with the positional scanning (Pinilla et al. 1992) approach resulted in the construction of a highly complex positional scanning combinatorial library consisting of 68,590 peptide mixtures (Schneider-Mergener et al. 1996, Kramer et al. 1997). This library $XXXX[B_1B_2B_3X_1X_2X_3]XXXX$ comprises ten sublibraries with the different hexapeptide cores $[XXB_1B_2B_3X]$, $[XB_1XB_2B_3X]$, $[XB_1B_2XB_3X]$, $[XB_1XXB_2B_3]$, $[XB_1XB_2XB_3]$, $[XB_1B_2XXB_3]$,

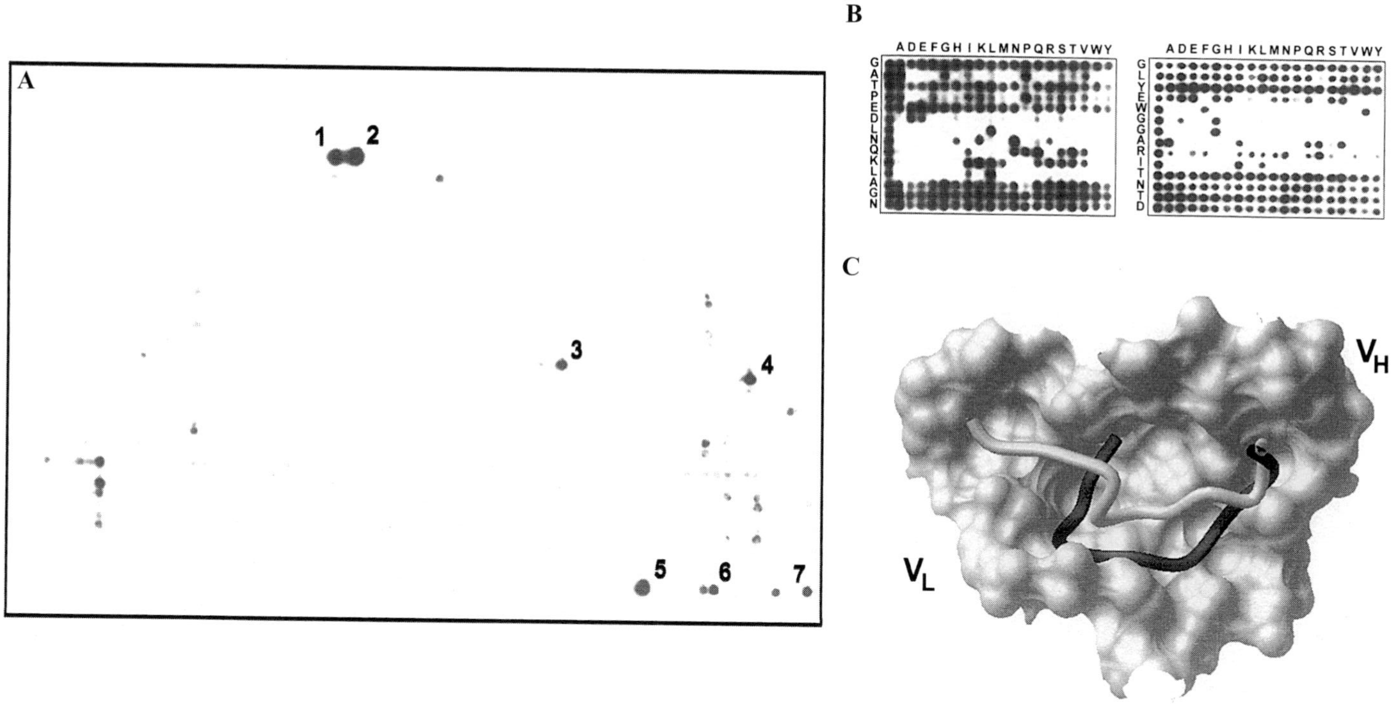


**Fig. 6A–C.** Identification of antibody CB4–1 binding peptides with a positional scanning combinatorial library. **A** Binding of the anti-p24 (HIV-1) antibody CB4-1 to the positional scanning library XXXX[$B_1B_2B_3X_1X_2X_3$]XXXX (68,590 spots in total) (Kramer et al. 1997; for description of the library see text). CB4-1 recognizes the p24 epitope GATPQDLNMTL (Höhne et al. 1993). From the complete library only a part (5,400 spots) is shown. The sequences of the spots 1–7 are: *1* XXXX [XNNXLX]XXXX, *2* XXXX[XNQXLX]XXXX, *3* XXXX[XKQXLX]XXXX, *4* XXXX [XLNXKX]XXXX, *5* XXXX[XGAXIX]XXXX, *6* XXXX[XGKXIX]XXXX and *7* XXXX [XGVXIX]XXXX. After iterative definition of the X-positions one epitope homologous peptide and three unrelated CB4-1 binding peptides were identified. **B** Substitutional analyses (see Fig. 3) of the epitope homologous peptide GATPEDLNQKLAGN and the unrelated peptide GLYEWGGARITNTD. **C** X-ray structure of the peptides GATPEDLNQKL (*light*) and GLYEWGGARIT (*dark*) in a complex with CB4-1 Keitel et al. 1997)

[$B_1XXXB_2B_3$], [$B_1XXB_2XB_3$], [$B_1XB_2XXB_3$] and [$B_1B_2XXXB_3$] corresponding to all possible distance patterns of three defined and undefined positions. Screening this library with the monoclonal anti-p24 (HIV-1) antibody CB4-1 resulted in the identification of the epitope GATPQDLNTML, which could not be detected with the libraries $XXB_1B_2XX$ or $XXXB_1B_2B_3XXX$ described above. More important, however, was that three completely unrelated peptide sequences were also obtained from this library (Fig. 6A, Kramer et al. 1997). Co-crystallization of the natural epitope as well as of one unrelated peptide (GLYEWGGARITNTD) demonstrated completely different binding conformations as well as interaction patterns for both peptides (Fig. 6, Keitel et al. 1997) helping explain the molecular basis for the binding promiscuity of this antibody. Searching databases for proteins containing the key interaction residues (supertopes, Kramer et al. 1997) of the identified peptides resulted in the identification of several CB4-1 binding proteins, including autoantigens. The immunological consequences of such polyspecific binding behavior by antibodies or T cells have been discussed as a possible mechanism for the onset of autoimmune diseases (Wucherpfennig and Strominger 1995).

## 6 Summary

The knowledge (antigen-derived peptide scans)- and library (de novo)-based mapping of linear and discontinuous antibody epitopes as well as protein-protein contact sites in general by spot synthesis now is a well established technique. Due to its automation, this technique also promises great potential for applications in functional genomics. It should help to elucidate the complex network of interacting protein molecules involved in signal transduction events (Adam-klages et al. 1996; Hoffmüller et al. 1999). Although little chemistry is involved in the preparation of peptide scans or libraries and the synthesis procedure is relatively simple, the laboratories of immunologists or molecular biologists are often not equipped to perform spot synthesis. In this case scans or libraries can be purchased from commercial suppliers.

*Acknowledgements.* This work was supported by grants of the Deutsche Forschungsgemeinschaft to U.R., Fonds der Chemischen Industrie to A.K. and the Charité. We wish to thank Avril Arhtur-Göttig for critical reading of the manuscript.

## References

Adam-Klages S, Adam D, Wiegmann K, Struve S, Kolanus W, Schneider-Mergener J, Krönke, M. (1996) The novel human TNF-receptor p55-associated WD-repeat protein FAN mediates neutral SMase activation. Cell 86:937–947

Cabilly S (ed) (1998) Combinatorial peptide library protocols. Meth Mol Biol 87, Humana Press, Totowa, USA

Cortese R (ed) (1996) Combinatorial libraries (synthesis, screening and application potential). Walter de Gruyter, Berlin/New York

Dong L, Schneider-Mergener J, Kramer A (1999) A novel type of amino acid property matrix based on functional studies of 68 continuous epitopes. Peptides 1998 (in press)

Duan Y, Laursen RA (1993) Protease substrate specificity mapping using membrane-bound peptides. Anal Biochem 216:431–438

Frank R (1992) Spot synthesis: An easy technique for the positionally addressable, parallel chemical synthesis on a membrane support. Tetrahedron 48:9217–9232

Frank R, Overwin H (1996) Spot synthesis: Epitope analysis with arrays of synthetic peptides prepared on cellulose membranes. Meth Mol Biol 66:149–169

Gao B, Esnouf MP (1996) Multiple interactive residues of recognition. J Immunol 157:183–188

Geysen HM, Meloen RH, Barteling SJ (1984) Use of peptide synthesis to probe viral antigens for epitopes to resolution of a single amino acid. Proc Natl Acad Sci USA 81:3998–4002

Geysen HM, Rodda SJ, Mason TJ (1986) A priori delineation of a peptide which mimics a discontinuous antigenic determinant. Mol Immunol 23:709–715

Geysen HM, Mason TJ, Rodda SJ (1988) Cognitive features of continuous antigenic determinants. J Mol Recognition 1:32–41

Hoeprich Jr PD, Langton BC, Zhang J-W Tam JP (1989) Identification of immunodominant regions of transforming growth factor α. J Biol Chem 264:19086–19091

Höhne WE, Küttner G, Kießig S, Hausdorf G, Grunow R, Winkler K, Wessner H, Gießmann E, Stigler R, Schneider-Mergener J, v.Baehr R, Schomburg D. (1993). Structural base of the interaction of a monoclonal antibody against p24 of HIV-1 with its peptide epitope. Mol Immunol 30:1213–1221

Hoffmann S, Frank R (1994) A new safety-catch peptide-resin linkage for the direct release of peptides into aqueous buffers. Tetrahedron Lett 35:7763–7766

Hoffmüller U, Russwurm M, Kleinjung F, Ashurst J, Vokmer-Engert R, Oschkinat H, Koesling D, Schneider-Mergener J (1999) Interactions of a PDZ domain with a synthetic library of all human protein C-termini. Angewandte Chemie Int Ed Engl (in press)

Haughten RA (1985) general method for the rapid aolid-phase synthesis of large numbers of peptides: Specificity of antigen-antibody interaction at the level of individual amino acids. Proc Natl. Acad Sci VSA 82: 5131–5135

Jones S, Thornton JM (1996) Principles of protein-protein interactions. Proc Natl Acad Sci USA 93: 13–20

Jung G (ed) (1996) Combinatorial peptide and nonpeptide libraries. VCH, Weinheim, Germany

Keitel T, Kramer A, Wessner H, Scholz C, Schneider-Mergener J, Höhne W (1997) Crystallographic analysis of anti-p24 (HIV-1) monoclonal antibody cross reactivity and polyspecificity. Cell 91: 811–820

Korth C, Stierli B, Streit P, Moser M, Schaller O, Fischer R, Schulz-Schaeffer W, Kretzschmar H, Raeber A, Braun U, Ehrensperger F, Hornemann S, Glockshuber R, Riek R, Billeter M, Wüthrich K, Oesch B (1997) Prion ($PrP^{Sc}$)-specific epitope defined by a monoclonal antibody. Nature 390:74–77

Kramer A, Volkmer-Engert R, Malin R, Reineke U, Schneider-Mergener J (1993) Simultaneous synthesis of peptide libraries on single resin and continuous membrane supports: identification of protein, metal and DNA binding peptide mixtures. Peptide Res 6:314–319

Kramer A, Schuster A, Reineke U, Malin R, Volkmer-Engert R, Landgraf C, Schneider-Mergener J (1994) Combinatorial cellulose-bound peptide libraries: screening tool for the identification of peptides that bind ligands with predefined specificity. METHODS (Comp Meth Enzymol) 6:388–395

Kramer A, Schneider-Mergener J (1995) Highly complex combinatorial cellulose-bound peptide libraries for the detection of antibody epitopes. In: Maia HLS (ed) Peptides 1994, Proceedings of the Twenty-Third European Peptide Symposium. ESCOM, Leiden, 475–476

Kramer A, Vacalopolou E, Schleuning WD, Schneider-Mergener J (1995) A general route to fingerprint analyses of peptide-antibody interactions using a clustered amino acid peptide library: comparison with a phage display library. Mol Immunol 32:459–465

Kramer A, Keitel T, Höhne W, Schneider-Mergener J (1997) Molecular basis for the binding promiscuity of an anti-p24 (HIV-1) monoclonal antibody. Cell 91:799–809

Kramer A, Schneider-Mergener J (1998) Synthesis and application of peptide libraries bound to continuous cellulose membranes. Meth Mol Biol 87:25–39

Kramer A, Affeldt M, Volkmer-Engert R, Schneider-Mergener J (1998) A novel type of protease cleavage assay based on cellulose-bound peptide libraries. In: Bajusz S, Hudesz S (eds) Peptides 1998, Proceedings of the Twenty-Fifth European Peptide Symposium. Mayflower Scientific Ltd., Kingswinford (in press)

Laune D, Molina F, Ferrieres G, Mani JC, Cohen P, Simon D, Bernardi T, Piechaczyk M, Pau B, Granier C (1997) Systematic exploration of the antigen binding activity of synthetic peptides isolated from the variable regions of immunoglobulins. J Biol Chem 272:30937–30944

Malin R, Steinbrecher A, Semmler W, Noll B, Johannsen B, Frömmel C, Höhne W, Schneider-Mergener J (1995) Identification of technetium-99 m binding peptides using cellulose-bound combinatorial peptide libraries. J Am Chem Soc 117:11821–11822

Molina F, Laune D, Gougat C, Pau B, Granier C (1996) Improved performances of spot multiple peptide synthesis. Pept Res 9:151–155

Morris GE (1996) Choosing a method for epitope mapping. In: Morris GE (ed) Epitope Mapping Protocols.Humana Press, Totowa NJ

Mukhija S, Germeroth L, Schneider-Mergener J, Erni B (1998) Identification of enzyme I inhibitors of the bacterial phototransferase system using combinatorial cellulose-bound peptide libraries. Eur J Biochem 254:433–438

Pavia MR, Sawyer TK, Moos WH (1993). The generation of molecular diversity. BioMed Chem Lett 3:387–396

Pinilla C, Appel JR, Blanc P, Houghten RA (1992) Rapid identification of high affinity peptide ligands using positional scanning synthetic peptide combinatorial libraries. BioTechniques 13:901–905

Pinilla C, Appel JR, Houghten RA (1993). Functional importance of amino acid residues making up peptide antigenic determinants. Mol. Immunol. 30:577–585

Reineke U, Sabat R, Kramer A, Stigler RD, Seifert M, Michel T, Volk HD, Schneider-Mergener J (1996) Mapping of protein-protein contact sites using cellulose-bound peptide scans. Mol Diversity 1: 141–148

Reineke U, Sabat R, Volk HD, Schneider-Mergener J (1998a) Mapping of the interleukin-10/interleukin-10 receptor combining site. Prot Sci 7:951–960

Reineke U, Ehrhard B, Sabat R, Volk HD, Schneider-Mergener J (1998b) Novel strategies for discontinuous epitope mapping using cellulose-bound peptide and hybritope scans. In: Ramage R, Epton R (eds) Peptides 1996, Proceedings of the Twenty-Fourth European Peptide Symposium. Mayflower Scientific Ltd., Kingswinford

Reineke U, Bhargava S, Schutkowski M, Landgraf C, Germeroth L, Fischer G, Schneider-Mergener J (1999) Spatial addressable fluorescence-quenched peptide libraries for the identification and characterization of protease substrates. In: Bajusz S, Hudesz S (eds) Peptides 1998, Proceedings of the Twenty-Fifth European Peptide Symposium. Mayflower Scientific Ltd., Kingswinford (in press)

Reuter M, Schneider-Mergener J, Kupper D, Meisel A, Mackeldanz P, Krüger D, Schroeder C (1999) ECORII DNA target recognition sites identified by membrane-bound peptide repertoires. J Biol Chem 274:5213–5221

Rüdiger S, Germeroth L, Schneider-Mergener J, Bukau B (1997) Substrate specificity of the DnaK chaperone determined by screening of cellulose-bound peptide libraries. EMBO J 16:1501–1507

Schneider-Mergener J, Kramer A, Reineke U (1996) Peptide libraries bound to continuous cellulose membranes: Tools to study molecular recognition. In: Cortese R (ed) Combinatorial Libraries. W. de Gruyter, Berlin

Strop P (1994). Synthetic combinatorial peptide libraries: a powerful tool for basic research and drug discovery. METHODS (Comp Meth Enzymol) 6:351–353

Tegge WJ, Frank R (1998) Analysis of protein kinase substrate specificity by the use of peptide libraries on cellulose paper (SPOT-method). Meth Mol Biol 87:99–106

Toomik R, Edlund M, Ek P, Obrink B, Engstrom L (1996) Simultaneously synthesized peptides on continuous cellulose membranes as substrates for protein kinases. Pept Res 9:6–11

Volkmer-Engert R, Hoffmann B, Schneider-Mergener J (1997) Stable attachment of the HMB linker to continuous cellulose membranes for parallel solid phase spot synthesis. Tetrahedron Lett 38:1029–1032

Weiergräber O, Schneider-Mergener J, Grötzinger J, Wollmer A, Köster A, Exner M, Heinrich PC (1996) Use of immobilized synthetic peptides for the identification of contact sites between human interleukin-6 and its receptor. FEBS Lett 379:122–126

Wucherpfennig KW, Strominger JL (1995) Molecular mimicry in T-cell-mediated autoimmunity: Viral peptides activate human T cell clones specific for myelin basic protein. Cell 80:695–705

Zdanov A, Schalk-Hihi C, Wlodawer A (1996) Crystal structure of human interleukin-10 at 1.6Å resolution and a model of a complex with its soluble receptor. Prot Sci 5:1955–1962

# The Two Hybrid Toolbox

W. Kolanus



## 1 Introduction

Protein–protein interactions are required for nearly every cellular function and are therefore universally studied in biology. Traditionally such interactions have been identified and characterized by the means of biochemistry, and the power of this approach is certainly undisputed. However, besides the fact that in the "old days" the composition of protein complexes was mostly studied in vitro, the arsenal of biochemical methods is very effective for tight interactions only, especially those with slow off-rate kinetics. Protein purification protocols have substantial time requirements, which means that interactions of interest might simply not survive the isolation procedure. Chemical cross-linking, i.e. covalent tethering of proteins, can sometimes overcome this limitation but this is not a widely used method. This is partly due to the fact that specific cross-linking of proteins may not be feasible under relevant conditions, e.g. inside a cell.

---

Laboratorium für Molekulare Biologie, Genzentrum der Universität München, Feodor-Lynen Str. 25, D-81377 Munich, Germany

The two hybrid system was designed to detect and to study protein–protein interactions in vivo. The original report was published 9 years ago, and it is clear that the technology has made a strong impact on the study of protein–protein interactions, and on biology in general. The reasons are obvious: the method is relatively easy to establish and it directly yields isolated genes. It is particularly useful in the study of proteins from organisms that are not well suited to either classical genetic or biochemical analyses (e.g. humans, drosophila). These features are also important aspects for the application of the two hybrid system in the context of functional genomics, and have led to the development of various related genetic screens. Here I will give an account on the uses of the classical two hybrid system and will further review some of the new hybrid technologies. It is important to mention that several excellent reviews have been published in recent years which cover a number of the aspects presented here in greater detail (Allen et al. 1995; Bartel and Fields 1995; Brachmann and Boeke 1997; Brent and Finley 1997; Fields and Sternglanz 1994; Frederickson 1998; Mendelsohn and Brent 1994; Warbrick 1997; White 1996).

## 2 The Classical Two Hybrid System

The essence of the two hybrid system is the modular architecture of a group of eukaryotic transcription factors, of which the yeast GAL4 transcription factor is the prototype (Ma and Ptashne 1987). These factors consist of a DNA binding domain (DB), which confers specificity for a given promoter sequence and a mostly acidic activation domain (AD) that interacts and activates the transcription machinery. Two important discoveries made the two hybrid system possible. First, the AD may be fused to a heterologous DB and still activate transcription of an appropriate promoter. Second, the two domains do not have to be covalently tethered to perform their function, i.e. it is possible to bridge them through a secondary interaction (Ma and Ptashne 1987). In the two hybrid system (Fig. 1), activation of reporter constructs occurs when the two domains are brought together through the interaction of two proteins expressed as AD- or DB-fusion proteins (Chien et al. 1991; Fields and Song 1989).

The implementation of the two hybrid system requires expression of two fusion proteins, usually encoded by independent vectors. One of them, the "bait," encodes an appropriate sequence specific DB, e.g. the one from GAL4 (Chien et al. 1991), or the bacterial repressor LexA (Gyuris et al. 1993). Usually, the DB is fused to the $NH_2$-terminal of the protein of interest. The second chimera, the "prey," encodes a transcription activation domain. The ADs which are in wide use are either derived from GAL4, or from the herpes simplex VP16 protein. As is the case for the DB fusion protein, the prospective interacting protein or the cDNA library is normally fused to the COOH-terminal of the AD. It should be noted that the AD may also be composed of artificial acidic sequences, which fortuitously activate transcription

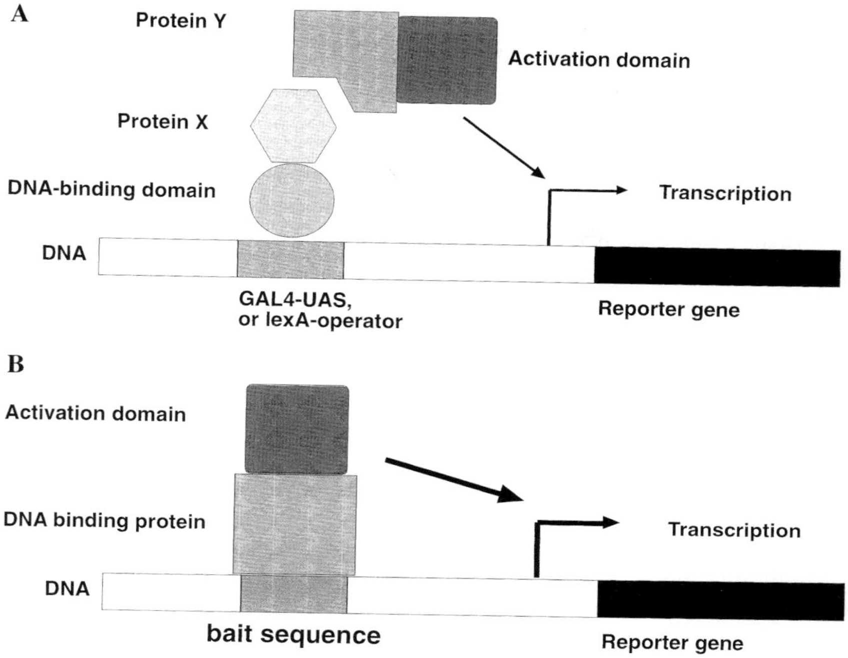


**Fig. 1. A** The two hybrid system: transcription of a reporter gene is activated through the interaction of two fusion proteins (bait and prey) which reconstitute a eukaryotic transcription factor, composed of DNA-binding and -activation domains. **B** The one hybrid system: an activation domain is fused to cDNA library sequences. In this simple setup the DNA sequence is the bait, and potential target proteins for this sequence are being screened for

when fused to a DB (GYURIS et al. 1993). The advantage of the artificial ADs is the fact that they are available in a wider range of biological activities, which means that the sensitivity of the two hybrid system can be fine-tuned with the help of these elements. It may for instance be that the enormous activation potential of, e.g. the VP16-AD, may lead to the amplification of non-specific interactions, or, conversely, to the "squelching-effect", in which the strongly overexpressed soluble AD chimera can compete for limiting cellular factors with the DNA-bound AD subpopulation and may thereby abrogate or attenuate reporter gene transcription.

The chimeric proteins are either expressed by strong constitutive promoters, such as ADH1, or from a conditional promoter (e.g. GAL1). Conditional promoters have the advantage of providing valuable criteria for the discrimination of false positives from bona fide interactors. Inducibility of one or of both fusion proteins may also aid in minimizing the potential toxic effects of a given interaction, because the time interval during which both proteins are expressed inside a given cell can be kept to a minimum.

The available two hybrid systems are mostly designed to be used in yeast, and therefore the third essential component is a yeast strain which must fulfill certain requirements, including auxotrophic mutations for the maintenance of the various plasmids involved, and, in the GAL4 based two hybrid system, deletion of genes encoding wild-type GAL4 as well as the GAL4 binding protein GAL80. The strain must further contain regulated reporter constructs which are either controlled by a lexA-operator or a GAL4-specific DNA element. These are integrated in the genome or present in the cell either as multicopy plasmids or as centromer-based single copy vectors. The precise layout of the reporter contributes to the sensitivity of the system, because number and affinity of the DNA elements as well as their distance to promoter core elements determines the relative strength of the cassette.

Commonly used reporters are yeast genes, such as *LEU2* or *HIS3*, which confer growth on minimal media to those cells in which an interaction occurred. Practically all systems further employ a second reporter, the *LacZ* gene, which encodes the enzyme β-galactosidase. The expression of β-galactosidase is visualized by a colorimetric assay, thus providing a strong second criterion for an occurring interaction.

Growth selection bears the inherent danger of suppression of toxic interactions, even when safeguarding steps like conditional expression of one of the two hybrids are employed (see above). If this is suspected, the possibility remains of doing color discrimination screening only. However, this is not done very frequently anymore, simply because it is a much more tedious procedure.

## 3 Performing a Two Hybrid Screen

Among the important reasons for the success of the method is the lack of sophisticated techniques involved, but basic yeast manipulation procedures should be firmly established in the lab. Whether or not a protein is suitable for a screening is very hard to predict. Often the bait protein has an inherent tendency to activate the reporter gene directly. Fortunately, a broad arsenal of methods now exists to get past this problem. These include low-sensitivity reporter systems, or counter-selection methods in which the reporter protein, e.g. *HIS3*, can be competitively inhibited by agents such as 3-aminotriazole. This allows a fine tuning of the reporter dose which hits the yeast cell, so the self activation potential of the bait can thus be greatly reduced. However, if a useful signal-to-noise ratio is not achieved by all these means, it is still possible to use truncated versions of the protein.

The outcome of a two hybrid screening is unpredictable, so one should be prepared for an uninformative result. Use of the two hybrid system for the analysis of single protein–protein interactions usually gives more straightforward data, but these will have to be backed up by biochemical experiments to be acceptable for publication. To put things in a realistic perspective it should be noted that it has become very difficult to publish two hybrid results without in vivo co-immunoprecipitation (co-IP) data for the interacting proteins. Although numerous weak interactions between cellular proteins are known which are nonetheless of physiological significance, it is understandable that co-IP data are required to validate a two hybrid result, simply to have a means of discriminating between important and more spurious interactions. Ironically, this partially leads to discarding one of the strong and unique aspects of the two hybrid system – the in vivo detection of high off-rate interactions, i.e. the ones you would normally not see by using immunoprecipitation techniques! I think that solid in vitro data, e.g. pull-down experiments in which one is able to titrate the interaction partners through a range of concentrations, in conjunction with indirect in vivo data which address cellular function, provide excellent arguments for the importance of any given interaction, even in the absence of co-IPs.

Negative two hybrid results may be due to a variety of reasons. These are, for example: protein instability, problems with nuclear import, and toxic effects to the cell. Some of these possibilities are relatively easily testable. The fusion proteins are usually very well expressed and their size and integrity can therefore by examined by western blotting. In the lexA system a repressor assay is available to assess whether a protein enters the nucleus. To this end, a reporter construct is used which expresses *LacZ* from the GAL1 promoter, but which contains a lex operator between the GAL1 upstream activation sequence and the core promoter. Binding of the LexA chimera results in a decrease in *LacZ* expression (Finley and Brent 1995). Another problem that has not really been explored in depth is steric inhibition, because the fusion proteins have to interact in an environment of complex protein–protein interactions, and these are all required for the proper functioning of the transcription machinery. Unfortunately, they are also not very well understood. It is therefore at any rate advisable to engineer peptide hinges (e.g., glycine or alanine spacers) into the chimeras which may provide the fusion partners with somewhat improved spatial flexibility.

A related aspect, and one that finds surprisingly little attention, is the inherent limitation imposed on any screen by the quality and the composition of the cDNA library used. If, for example, the hypothetical interacting domain you are hunting for resides at the $NH_2$-terminal of a 250kDa protein (i.e., is encoded by a message of at least 7kb length), it is fairly obvious that this one will be much harder to find than the average protein. This is because, even if the library has been size-selected, it is clear that such a sequence will be severely underrepresented, and in case it is a non-abundant message the likelihood of not getting it almost equals "1". It is not surprising that many proteins yielded by two hybrid experiments are encoded by relatively abundant mRNAs, simply because their interactions with the bait proteins have been statistically favored. To avoid all of this, and to give rare interactions a chance to peek through, it would be advisable to use normalized libraries, i.e. ones in which all cDNAs have roughly equal distributions. Furthermore, to steer around the problem of having to isolate very large cDNAs, it is recommended to use randomly primed libraries, rather than oligo-dT primed libraries. Those would not have to be painstakingly size-selected; on the contrary one would create "domain" libraries. Of course this also means that the chance of pulling out a full-length clone in one step is decreased. It should also be noted that it is more difficult to prepare a very good randomly primed library, primarily because the making of one crucially depends on very high-quality mRNA. If the mRNA isolation and purification step is not performed in a rigorous fashion, the library will be heavily contaminated by ribosomal RNA. This is much less of a problem when mRNA-specific primers are used, and this is why we mostly see oligo-dT primed libraries around.

A very common problem associated with the two hybrid technique is the identification of false positives. These are fusion proteins which show a high level of nonspecific interactions. A table of commonly found false positives has been compiled (http://www.fccc.edu:80/research/labs/golemis/InteractionTrapInWork). The elimination of false positives requires control experiments with a wide variety of control baits. Sometimes interactions occur because the two hybrid system forces two proteins to the same environment of the yeast nucleus, whereas they would normally be sequestered. Strong nonspecific "stickiness" of a bait is not that common, but the two hybrid system does indeed pick up many interactions which unfortunately do not help immediately to address the physiological role of the bait (i.e. interactions with ribosomal proteins, with chaperones, or with proteins of the ubiquitination machinery). However, it is entirely possible that some of these interactions may actually take place in certain cellular situations, so one should be careful with the term "false positive".

Essentially all methods for analyzing protein interactions can result in artifacts. On the other hand, hardly any other approach has yielded such a massive proliferation of candidate interactors for so many different proteins in such a short period of time. So, weeding out the "bad" ones early remains a very important issue.

## 4 Analysis of Protein Interaction Domains

The two hybrid system has been widely and very successfully used to detect novel protein–protein interactions. It bears the advantage that the interactions may be detected inside a eukaryotic cell and many interactions can be rapidly and simultaneously screened. Although the yeast nucleus is an exotic environment to many of the interactions which are tested with this method, it seems likely that this may sometimes even be advantageous: homologous cellular factors which might compete with the interactors or otherwise perturb the reaction are simply not expressed in this compartment! In this context it is noteworthy that the procedure has been particularly useful in the isolation of interactors for cytoplasmic domains of transmembrane proteins (Adam et al. 1996; Ansieau et al. 1996; Bressler et al. 1996; Campbell et al. 1995; Campbell and Giorda 1997; Chen et al. 1995; Daram et al. 1997; Eigenthaler et al. 1997; Guenette et al. 1996; Hu et al. 1994; Kolanus et al. 1996; Lee et al. 1996; Margottin et al. 1996; Mcloughlin and Miller 1996; Meyer et al. 1997; Naik et al. 1997; Rothe et al. 1994; Song and Donner 1995).

The two hybrid method directly yields information about the primary structure of the interacting protein and isolation of truncated preys may even result in an immediate rough mapping of the interaction domains. Since the level of reporter activation correlates well with the specific binding activities, measurement of reporter activity gives an indication of the strength of the interaction (ESTOJAK et al. 1995).

The binding of bait proteins to wild type and mutant versions of the prey can be compared. IWABUCHI et al. (1994) have used this method to identify proteins that bind to wild type, but not mutant forms of p53. By contrast, a functional point mutant of the bait may provide a powerful criterion to select for interactors which are located downstream of the bait protein in a specific cellular pathway. STANGER et al. (1995) have identified the RIP protein as a cytoplasmic ligand for the FAS receptor. The FAS system is involved in the regulation of programmed cell death. RIP was characterized by an important feature which separated it from the pool of primary interactors: it lacked the ability to bind to the *lpr* mutant of the intracellular tail of FAS. This mutation had previously been implicated in a severe dysfunction of the immune system.

Once an interaction between two proteins in the two hybrid system has been established, then specific interaction domains can be investigated. Subdomains of proteins may be expressed, or deleted, and the resulting constructs may be tested for interactions with known partners; it is thus possible to identify distinct interaction domains. For example, this approach has been successful in defining the region of proliferating cell nuclear antigen (PCNA) essential for its interaction with the CDK inhibitor $p21^{CIP1}$, and these results have been confirmed by structural studies (WARBRICK et al. 1995). Studies of homodimerization can very conveniently be carried out in the two hybrid system since labeling or tagging of proteins is not required. CHIEN et al. (1991) extensively studied homodimerization of the SIR4 protein and identified the small region required for this process. Another advantage of the system lies in the genetic nature of the two hybrid approach: mutant plasmids can be screened for changes that disrupt the protein–protein interaction by looking for loss of reporter expression (also see below). Such an approach has been used to screen for p53 mutants which can no longer bind to SV40 large T antigen (LI and FIELDS 1993). REYMOND and BRENT (1995) have used the two hybrid system in a study of the cyclin-dependent kinase (CDK) binding capacity of pl6 allelic variants detected in melanoma-prone families. The same group has recently developed a two-bait two hybrid-system which enables them to detect a set of logical relationships among proteins. This directly leads to the possibility of performing model examinations of the functions of complex protein networks with the help of a relatively simple system, i.e. "bio-cybernetics" (XU et al. 1997).

INOUYE et al. (1997) have developed a variant of the two hybrid system which permits a mutational analysis of the interaction of a protein with one or another of two separate partners. They screened mutants of the Ste5 protein, expressed as AD fusion proteins, for effects on their interactions with Ste7 and Ste11. The latter were fused to distinct DBs (Gal4 or LexA) which bound to separate reporter genes, *URA3* and *lacZ*, respectively, allowing selection for mutants affecting either the Ste5-Ste7 or the Ste5-Ste11 interaction.

## 5 Variations of the Original: Reverse and Counterselection Two Hybrid Systems

It is often desirable to screen for the disruption of a given interaction, e.g. in order to create loss-of-function mutants of the bait/prey proteins or to identify specific inhibitors of their interaction. This is strongly demanded for industrial scale high-throughput screenings with the objective of targeting protein–protein interactions of clinical or commercial interest. There are numerous efforts to adapt the system to the purpose lead-substance identification, but to this end the use of the classical two hybrid system coupled to a negative read-out would probably be too artifact-prone.

Positive selection schemes have therefore been devised to facilitate the identification of mutations, proteins, peptides or drugs that disrupt two hybrid interactions. Vidal et al. (1996) (Fig. 2) replaced the standard two hybrid *HIS3* reporter gene with the counterselectable *URA3*. The *URA3*-encoded enzyme is toxic to cells grown on 5-fluoroorotic acid, and thus interaction of the activation domain and DNA-binding domain fusion proteins results in sensitivity to this compound. Conversely, dissociation of the two hybrid interaction results in loss of *URA3* expression, 5-fluoroorotic acid resistance and cell growth. A similar result was achieved with the so-called split-hybrid system (Shih et al. 1996), substituting the *Escherichia coli* transcriptional repressor TetR for the two hybrid *HIS3* reporter (Fig. 3). A two hybrid interaction results in expression of TetR which binds and inhibits expression of a *HIS3* reporter which had been engineered to contain a TetR binding site; disruption of the two hybrid pair restores *HIS3* expression and provides a selective growth advantage. The inherent advantage of both these systems is that they greatly facilitate the selection for rare mutations or compounds that disrupt an interaction of interest by employing a positive selection process. Dixon et al. (1997) have devised a similar selection screen to enable the screening for proteases of a defined sequence specificity in a modified mammalian two hybrid screen.

## 6 Interactions Involving Three Elements

Interactions which are dependent on post-translational modifications that do not commonly occur in yeast, such as protein tyrosine phosphorylation, cannot be studied with the two hybrid system. To overcome this limitation, Osborne et al. (1995) coexpressed in yeast the T cell-specific tyrosine kinase lck, capable of phosphorylating a bait hybrid, the high affinity Fcε receptor, on specific target residues to facilitate its interaction with the SH2 domains of the Syk protein, which was expressed as an AD hybrid. As expected, the interaction of Syk with the Fcε receptor was only detected in the presence of the lck protein (Osborne et al. 1995). This approach is not limited to the study of protein phosphorylation as it should in

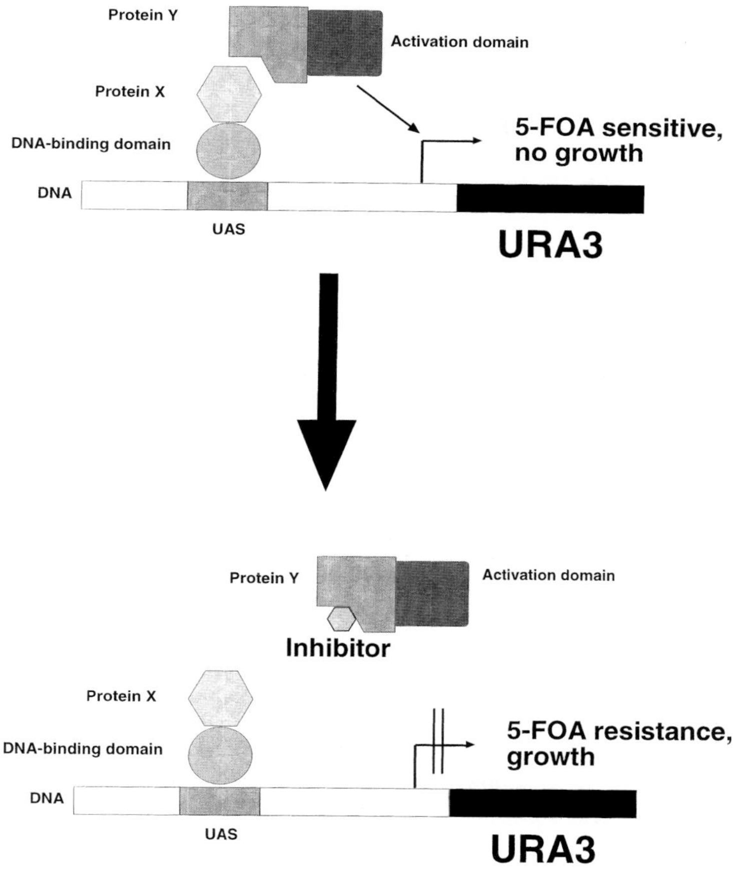


**Fig. 2.** The reverse two hybrid system, variant 1: interaction of two fusion proteins leads to transcription of the URA3 gene, the product of which is toxic if the cells are grown on 5-FOA. Expression of an inhibitor leads to 5-FOA resistance and therefore to cell growth

principle be possible to provide other modifying factors to facilitate analysis of post-translational or allosteric regulations (Fig. 4A). A possible future application may be to use substrate trapping mutants of an enzyme (e.g. protein phosphatase) as baits to screen for bona fide in vivo substrates (Flint et al. 1997). Alternatively, expression of a third protein may be used to bridge, stabilize, or inhibit a two hybrid interaction, allowing further experimental variation (see the section on three hybrid systems below). The use of mammalian two hybrid systems is an alternative when post-translational modifications may be a concern. However, these systems are not yet amenable to large scale screening.

Membrane-anchored proteins often contain signals that interfere with the nuclear localization necessary for two hybrid reporter expression. If this is known, such signal sequences may be removed from the bait protein (e.g. myristoylation or

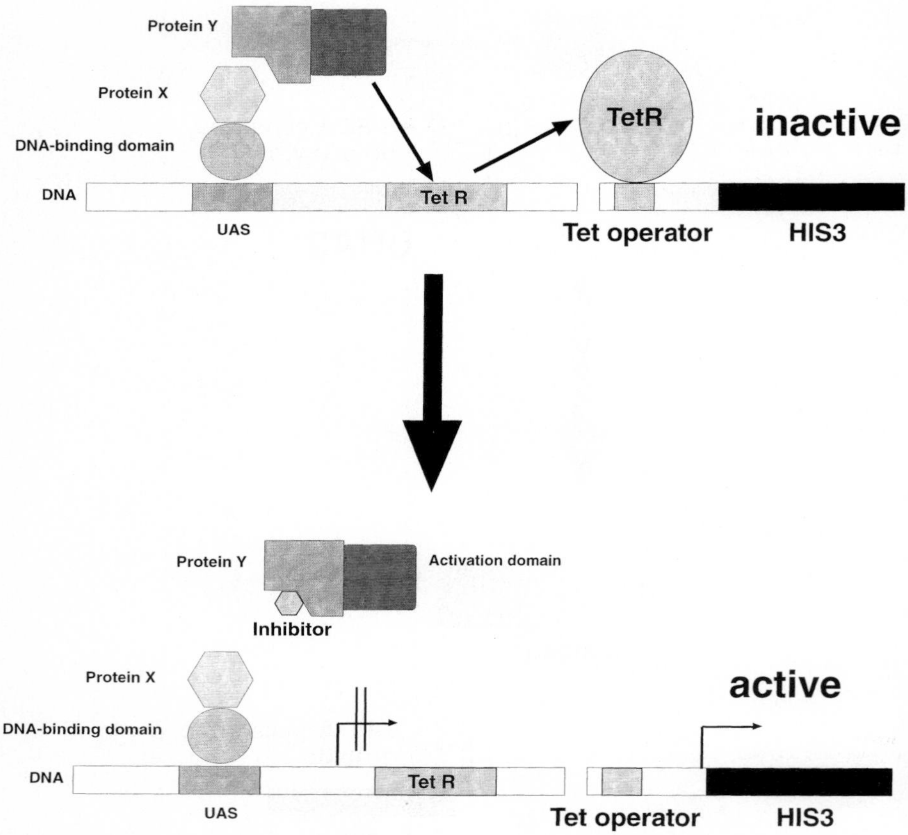


**Fig. 3.** Reverse two hybrid system, variant 2: a protein-protein interaction triggers the expression of the Tet repressor which in turn blocks the activation of a reporter gene. Inhibition of the interaction therefore leads to reporter gene activation

farnesylation signals). On the other hand, some proteins require membrane localization for optimal interaction with a partner. Post-translational modifications, which may take place in the secretory pathway or in the extracellular environment, cannot occur in the nucleus. These are, for example, glycosylation and/or disulfide bond formation. Although extracellular receptor-ligand combinations have been detected with the help of the two hybrid system (Ozenberger and Young 1995), these are the exception rather than the rule, and alternative systems to facilitate the study of such interactions are urgently needed.

## 6.1 Peptide Ligand Three Hybrid System

Studies with extracellular domains of transmembrane receptor tyrosine kinases gave rise to a peptide ligand-based three hybrid system (Ozenberger and Young

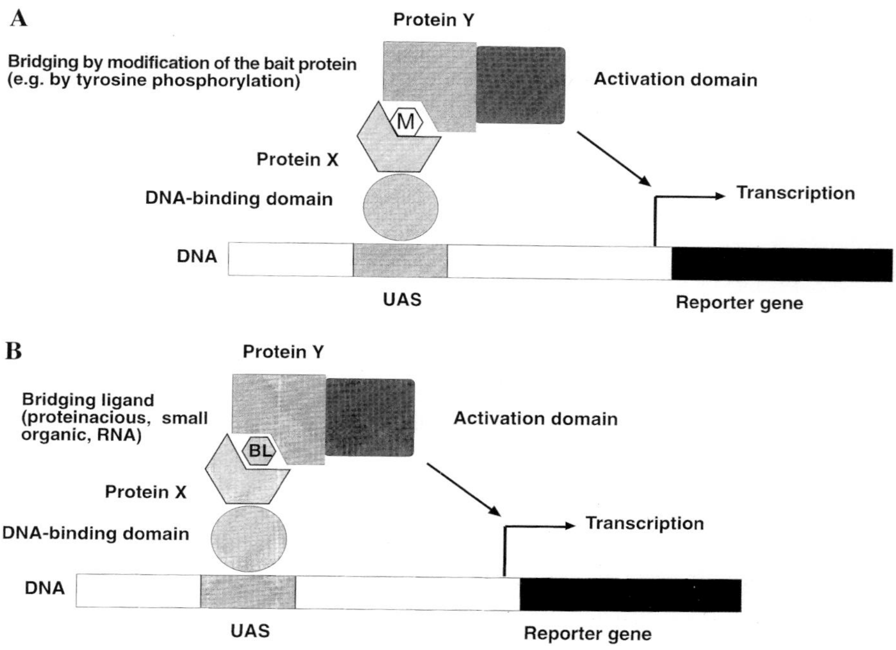


**Fig. 4. A** Three ligand two hybrid system, part 1: the interaction between the hybrid proteins is dependent on bait protein modification which must be provided in *trans*. **B** Three ligand two hybrid system, part 2: a bridging ligand is required to bring the two hybrid proteins together. This may be a protein ligand, a small organic ligand or an RNA molecule. The bridging ligand needs to have binding surfaces for both fusion proteins

1995). The extracellular domain of a transmembrane receptor is fused to the DB and separately to the Gal4 AD. Expression of the native peptide ligand for the receptor leads to receptor dimerization via peptide ligand binding and a transcriptional readout (Fig. 4B). This was shown for both the growth hormone receptor and the receptor for vascular endothelial growth factor, flk1/KDR (Ozenberger and Young 1995).

## 6.2 Small Organic Ligand Three Hybrid System

The small ligand three hybrid system (Licitra and Liu 1996) makes use of so-called synthetic chemical inducers of dimerization (CIDs) (Belshaw et al. 1996; Ho et al. 1996; Holsinger et al. 1995; Spencer et al. 1993, 1996) and has as its essential feature a heterodimer of covalently linked small organic ligands, for example, dexamethasone and FK506 (Fig. 4B). Dexamethasone interacts with a rat glucocorticoid receptor-LexA DB hybrid and FK506 interacts with its binding protein, FKBP12, fused to an AD. The small organic hybrid ligand thus bridges the two chimeric proteins, thereby leading to reporter activation. A cDNA encoding

FKBP12 was recovered in a library screen performed in the presence of the small hybrid ligand. The CID technology has been used in mammalian and in yeast experiments. Its combination with the two hybrid system will likely have a variety of practical applications. However, since the CID system depends on diffusion of the hybrid ligand into yeast cells, permeability problems may also limit its applicability.

## 6.3 RNA Three Hybrid System

The third ligand may also be a nucleic acid: the RNA three hybrid system requires two hybrid proteins and one hybrid RNA (Fig. 4B). One part of the hybrid RNA is required for the constant, known interaction, whereas the other part (the "bait RNA") is used to screen for so-called orphan RNA-binding proteins. Putz et al. (1996) have fused the HIV-1 Rev M10 mutant to the Gal4 DB. This Rev mutant binds tightly to its target RNA, the Rev responsive element (RRE). This three hybrid system will identify new RNAs recognized by an orphan RNA binding protein fused to the Gal4 AD, if a library of RRE hybrid RNAs is utilized. A variation of this principle also allows for the selection for novel RNA-binding proteins.

Sengupta et al. (1996) fused an RNA-phage capsid protein as the bait RNA-binding protein to the LexA DB. The capsid protein recognizes a 21-nucleotide RNA stem-loop in the phage genome with high affinity, and this feature was used to construct hybrid RNAs. Both this protein-RNA interaction, as well as that of HIV-1 Tat protein and its RNA target, TAR, were shown to be valid in vivo partners in this RNA three hybrid system. This system has been successfully applied to identify a novel histone RNA 3′ hairpin binding protein (Martin et al. 1997; Wang et al. 1996).

# 7 Extensions of the Two Hybrid Approach

An important limitation to the original system is that bona fide or artificial Pol II transcriptional activators cannot generally be used as baits, due to their tendency to stimulate reporter expression in the absence of a two hybrid interaction. To get around this shortcoming, Aronheim et al. (1994, 1997) developed a system that operates in the yeast cytoplasm (Fig. 5) and relies on the capacity of the human guanyl nucleotide exchange factor (GEF) hSos to substitute for a temperature-sensitive mutant of the yeast Ras GEF, Cdc25. HSos must be recruited to the cell membrane where it can stimulate guanyl nucleotide exchange on yeast Ras. Interaction of a cytoplasmic hSos-bait protein fusion with a protein partner targeted to the cell membrane through fusion to the Src myristoylation signal provides GEF activity and permits cell survival at the non-permissive temperature.

ARONHEIM et al. (1997) isolated two new binding proteins for c-Jun, JDP1 and JDP2 and thereby demonstrated that this system can be used in library screens.

One shortcoming of the system was the reported background of isolated suppressor mutants which activate the Ras pathway downstream of Sos (BALLESTER et al. 1989). However, recent advances have overcome this limitation: firstly, co-expression of mammalian GAP suppresses the isolation of ras encoding cDNAs during a SOS two hybrid screen (ARONHEIM 1997), and secondly, an improved system based on the direct recruitment of Ras to the plasma membrane circumvents this problem altogether (BRODER et al 1998).

In a different approach, MARSOLIER et al. (1997) created a two hybrid system making use of an RNA polymerase III transcription unit (Fig. 6). The B-box element within the promoter of a Pol III reporter gene *SNR6* was substituted with a binding site for the bait protein fusion. Interaction of X with protein partner Y fused to the t38 subunit of the Pol III transcription factor TFIIIC reconstitutes TFIIIC activity and leads to reporter gene expression. In this system, the *SNR6* gene itself, an essential component of the spliceosome, is used as a reporter. This is normally done in conjunction with a 5-FOA based procedure, in which the

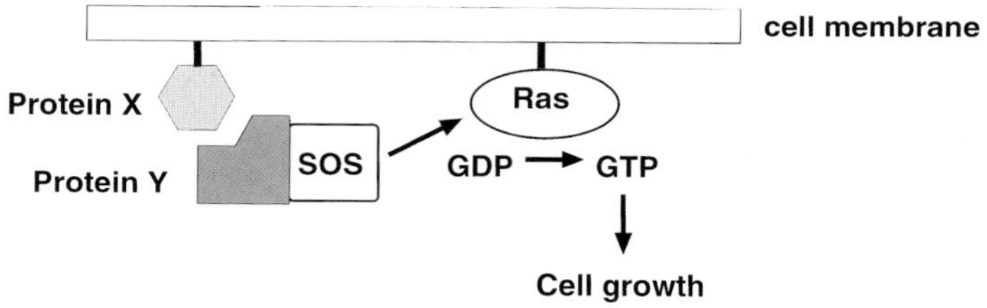


**Fig. 5.** The Sos two hybrid system: the bait protein is targeted to the plasma membrane of a yeast cell, whereas the interacting protein is fused to the hSos protein, a guanyl nucleotide exchange factor for Ras. Interaction of the fusion proteins results in membrane recruitment of the exchange factor, and thus to cell activation and growth

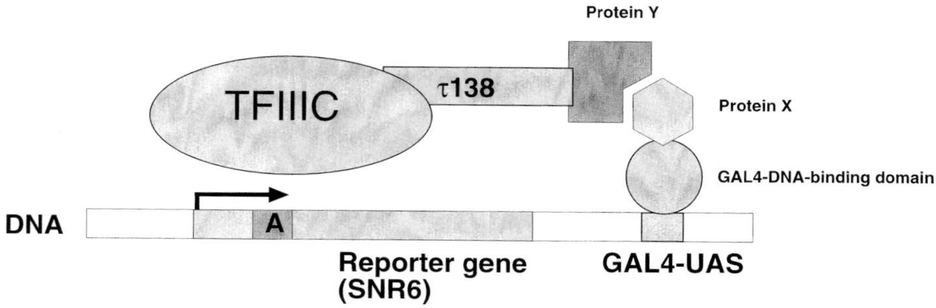


**Fig. 6.** The Pol III two hybrid system: the B-box element which is essential for transcription from a Pol III promoter is replaced by a GAL4-UAS, to which the bait protein can bind. The prey consists of an interactor for the bait fused to the $\tau$138 subunit of the Pol III transcription factor TFIIIC. Binding of the $\tau$138 in the vicinity of a Pol III promoter recruits the TFIIIC protein and thereby drives the reporter gene

expression of wild-type SNR6 from a *URA3* plasmid is counterselected (Marsolier et al. 1997).

The ubiquitin-based split-protein sensor (USPS) system developed by Johnsson and Varshavsky (Fig. 7) is based on the fact that newly formed fusions between ubiquitin and proteins are rapidly cleaved by ubiquitin-specific proteases (UBPs) in eukaryotes (Johnsson and Varshavsky 1994). This cleavage only occurs if ubiquitin is properly folded. Johnsson and Varshavsky demonstrated that the $NH_2$- and COOH-terminal portions of ubiquitin, when coexpressed as separate sub-entities, could still fold correctly. The readout for the experiment was a western blot analysis of a reporter protein, the hemagglutinin-tagged dihydrofolate reductase, which was proteolytically released from the COOH-terminal ubiquitin fragment by UBPs following assembly of both ubiquitin moieties. This cofolding of fragments did not occur in the presence of a particular $NH_2$-terminal missense mutant, probably by decreasing the affinity of the separate portions. However, when the two moieties were fused to known interacting proteins, the effect of the $NH_2$-terminal mutation was overcome, leading to correct assembly and to cleavage of the reporter protein from the COOH-terminal ubiquitin fragment. A less labor-intensive readout has recently been developed to make this highly original system amenable to large scale screening (Stagljar et al. 1998). It is based on the UBP-mediated cleavage of a transcription factor which subsequently activates a reporter gene in the nucleus.

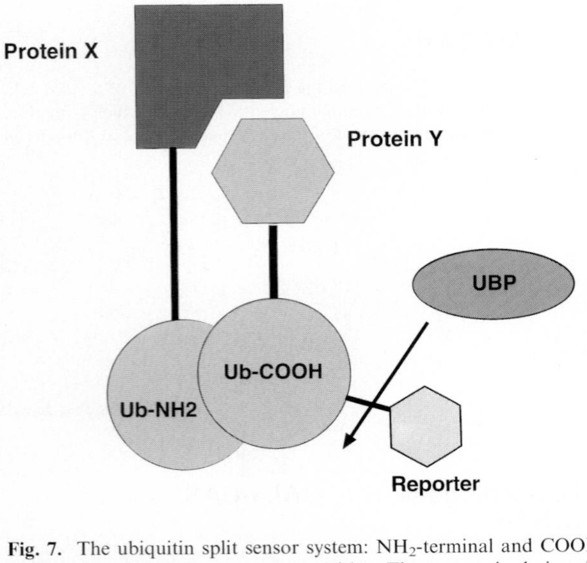


**Fig. 7.** The ubiquitin split sensor system: $NH_2$-terminal and COOH-terminal moieties of the ubiquitin protein are expressed as separate entities. The system is designed to allow the reconstitution of the ubiquitin protein from the separate portions only when a secondary interaction (in this case between proteins X and Y) assists in the assembly. Reconstitution of ubiquitin directs cleavage of a reporter moiety by a ubiquitin-dependent protease (UBP)

# 8 Two Hybrid Technology and Functional Proteomics: Protein Linkage Maps

Finally, it is well worth mentioning that tremendous efforts are directed at the application of the two hybrid system to screen for all possible protein interactions in cells or tissues of interest. The start has been made with the analysis of all-protein encoding genes, or the proteome, of yeast *Saccharomyces cervisae*. It is thought that new connections for proteins of known function will arise and that this approach will furthermore reveal the roles of many genes which are poorly understood. Smaller scale protein linkage maps have been successfully completed for bacteriophage T7 and for subsets of eukaryotic cellular proteins (Evangelista et al. 1996). However, the scaling of the two hybrid screen to permit systematic screening of the thousands of yeast proteins, or even the 100,000 proteins of the human genome, requires strong improvements in all areas of data acquisition and management. Iterative approaches are being used to meet these objectives (Cho et al. 1998; Fromont et al. 1997), and it is likely that in the future they will be combined with systems which utilize pre-organized arrays of vast numbers of proteins (Hudson et al. 1997). Important goals are automatic processing and miniaturizing of these arrays, while keeping a high discriminatory potential at the same time. High-density oligonucleotide array technology (the "DNA-chip") is now being widely applied to genome analyses. It is clear that an analogous device would likewise revolutionize the world of proteomics: it is unfortunately not available – at present.

## References

Adam KS, Adam D, Wiegmann K, Struve S, Kolanus W, Schneider MJ, Kronke, M (1996) FAN, a novel WD-repeat protein, couples the p55 TNF-receptor to neutral sphingomyelinase. Cell 86:937–947

Allen JB, Walberg MW, Edwards MC, Elledge SJ (1995) Finding prospective partners in the library: the two hybrid system and phage display find a match. Trends Biochem Sci 20:511–516

Ansieau S, Scheffrahn I, Mosialos G, Brand H, Duyster J, Kaye K, Harada J, Dougall B, Hubinger G, Kieff E, Herrmann F, Leutz A, Gruss HJ (1996) Tumor necrosis factor receptor-associated factor (TRAF)-1 TRAF-2 and TRAF-3 interact in vivo with the CD30 cytoplasmic domain; TRAF-2 mediates CD30-induced nuclear factor kappa B activation. Proc Natl Acad Sci USA 93:14053–14058

Aronheim A, Engelberg D, Li N, al-Alawi N, Schlessinger J, Karin M (1994) Membrane targeting of the nucleotide exchange factor Sos is sufficient for activating the Ras signaling pathway. Cell 78:949–961

Aronheim A, Zandi E, Hennemann H, Elledge S, Karin M (1997) Isolation of an AP-1 repressor by a novel method for detecting protein–protein interactions. Mol Cell Biol 17:3094–3102

Aronheim A (1997) Improved efficiency Sos recruitment system: expression of the mammalian GAP reduces isolation of Ras GTPase false positives. Nucl Acids Res 25:3373–3374

Ballester R, Michael T, Ferguson K, Xu H, Cormick FM, Wigler M (1989) Genetic analysis of mammalian GAP expressed in yeast. Cell 59:681–686

Bartel PL, Fields S (1995) Analyzing protein–protein interactions using two hybrid system. Methods Enzymol 254:241–263

Belshaw PJ, Ho SN, Crabtree GR, Schreiber SL (1996) Controlling protein association and subcellular localization with a synthetic ligand that induces heterodimerization of proteins. Proc Natl Acad Sci USA 93:4604–4607

Brachmann R, Boeke J (1997) Tag games in yeast: the two hybrid system and beyond. Curr Opin Biotechnol 8:561–568
Brent R, Finley RJ (1997) Understanding gene and allele function with two hybrid methods. Annu Rev Genet 31:663–704
Bressler SL, Gray MD, Sopher BL, Hu Q, Hearn MG, Pham DG, Dinulos MB, Fukuchi K, Sisodia SS, Miller MA, Disteche CM, Martin GM (1996) cDNA cloning and chromosome mapping of the human Fe65 gene: interaction of the conserved cytoplasmic domains of the human beta-amyloid precursor protein and its homologues with the mouse Fe65 protein. Hum Mol Genet 5:1589–1598
Broder YC, Katz S, Aronheim A (1998) The Ras recruitment system a novel approach to study protein–protein interactions. Curr Biology 8:1121–1124
Campbell KS, Buder A, Deuschle U (1995) Interactions between the amino-terminal domain of p56lck and cytoplasmic domains of CD4 and CD8 alpha in yeast. Eur J Immunol 25:2408–2412
Campbell K, Giorda R (1997) The cytoplasmic domain of rat NKR-P1 receptor interacts with the N-terminal domain of p56(lck) via cysteine residues. Eur J Immunol 27:72–77
Chen RH, Moses HL, Maruoka EM, Derynck R, Kawabata M (1995) Phosphorylation-dependent interaction of the cytoplasmic domains of the type I and type II transforming growth factor-beta receptors. J Biol Chem 270:12235–12241
Chien CT, Bartel PL, Sternglanz R, Fields S (1991) The two hybrid system: a method to identify and clone genes for proteins that interact with a protein of interest. Proc Natl Acad Sci USA 88:9578–82
Cho R, Fromont-Racine M, Wodicka L, Feierbach B, Stearns T, Legrain P, Lockhart D, Davis R (1998) Parallel analysis of genetic selections using whole genome oligonucleotide arrays. Proc Natl Acad Sci USA 95:3752–3757
Daram P, Urbach S, Gaymard F, Sentenac H, Cherel I (1997) Tetramerization of the AKT1 plant potassium channel involves its C-terminal cytoplasmic domain. Embo J 16:3455–3463
Dixon EP, Johnstone EM, Liu X, Little SP (1997) An inverse mammalian two hybrid system for beta secretase and other proteases. Anal Biochem 249:239–241
Eigenthaler M, Hofferer L, Shattil SJ, Ginsberg MH (1997) A conserved sequence motif in the integrin beta3 cytoplasmic domain is required for its specific interaction with beta3-endonexin. J Biol Chem 272:7693–7698
Estojak J, Brent R, Golemis EA (1995) Correlation of two hybrid affinity data with in vitro measurements. Mol Cell Biol 15:5820–5829
Evangelista C, Lockshon D, Fields S (1996) The yeast two hybrid system: prospects for protein linkage maps. Trends Cell Biol 6:196–199
Fields S, Song O (1989) A novel genetic system to detect protein–protein interactions. Nature 340:245–246
Fields S, Sternglanz R (1994) The two hybrid system: an assay for protein–protein interactions. Trends Genet 10:286–292
Finley R, Brent R (1995) Interaction trap cloning with yeast. In: B Hames, D Glover (eds) DNA cloning 2; expression systems: a practical approach. Oxford University Press, Oxford UK, pp 169–203
Flint A, Tiganis T, Barford D, Tonks N (1997) Development of substrate-trapping mutants to identify physiological substrates of protein tyrosine phosphatases. Proc Natl Acad Sci USA 94:1680–1685
Frederickson R (1998) Macromolecular matchmaking: advances in two hybrid system technologies. Curr Opin Biotechnol 9:90–96
Fromont RM, Rain JC, Legrain P (1997) Toward a functional analysis of the yeast genome through exhaustive two hybrid screens [see comments]. Nat Genet 16:277–282
Guenette SY, Chen J, Jondro PD, Tanzi RE (1996) Association of a novel human FE65-like protein with the cytoplasmic domain of the beta-amyloid precursor protein. Proc Natl Acad Sci USA 93:10832–10837
Gyuris J, Golemis E, Chertkov H, Brent R (1993) Cdi1 a human G1 and S phase protein phosphatase that associates with Cdk2. Cell 75:791–803
Ho SN, Biggar SR, Spencer DM, Schreiber SL, Crabtree GR (1996) Dimeric ligands define a role for transcriptional activation domains in reinitiation. Nature 382:822–826
Holsinger LJ, Spencer DM, Austin DJ, Schreiber SL, Crabtree GR (1995) Signal transduction in T lymphocytes using a conditional allele of Sos. Proc Natl Acad Sci USA 92:9810–9814
Hu HM, O'Rourke K, Boguski MS, Dixit VM (1994) A novel RING finger protein interacts with the cytoplasmic domain of CD40. J Biol Chem 269:30069–30072
Hudson J, Dawson E, Rushing L, Jackson C, Lockshon D, Conover D, Landicault C, Harris J, Simmons S, Rothstein R, Fields S (1997) The complete set of predicted genes from Saccharomyces cerevisiae in a readily usable form. Genome Res 12:1169–1173

Inouye C, Dhilon N, Durfree T, Zambryski P, Thorner J (1997) Mutational analysis of STE5 in the yeast Saccharomyces cerevisiae: application of a differential interaction trap assay foe examining protein–protein interactions. Genetics 147:479–492

Iwabuchi K, Bartel PL, Li B, Marraccino R, Fields S (1994) Two cellular proteins that bind to wild-type but not mutant p53. Proc Natl Acad Sci USA 91:6098–6102

Johnsson N, Varshavsky A (1994) Split ubiquitin as a sensor of protein interaction in vivo. Proc Natl Acad Sci 91:10340–10344

Kolanus W, Nagel W, Schiller B, Zeitlmann L, Godar S, Stockinger H, Seed B (1996) AlphaLBeta2 integrin/LFA-1 binding to ICAM-1 induced by a cytoplasmic regulatory molecule. Cell 86:233–242

Lee SY, Park CG, Choi Y (1996) T cell receptor-dependent cell death of T cell hybridomas mediated by the CD30 cytoplasmic domain in association with tumor necrosis factor receptor-associated factors. J Exp Med 183:669–674

Li B, Fields S (1993) Identification of mutations in p53 that affect its binding to SV40 large T antigen by using the yeast two hybrid system. Faseb J 7:957–963

Licitra EJ, Liu JO (1996) A three-hybrid system for detecting small ligand-protein receptor interactions. Proc Natl Acad Sci USA 93:12817–12821

Ma J, Ptashne M (1987) A new class of transcriptional activators. Cell: 51:113–119

Margottin F, Benichou S, Durand H, Richard V, Liu LX, Gomas E, Benarous R (1996) Interaction between the cytoplasmic domains of HIV-1 Vpu and CD4: role of Vpu residues involved in CD4 interaction and in vitro CD4 degradation. Virology 223:381–386

Marsolier MC, Prioleau MN, Sentenac A (1997) A RNA polymerase III-based two hybrid system to study RNA polymerase II transcriptional regulators. J Mol Biol 268:243–249

Martin F, Schaller A, Egite S, Schumperli D, Muller B (1997) The gene for histone RNA hairpin binding protein is located on human chromosome 4 and encodes a novel type of RNA binding protein. EMBO J 16:769–778

McLoughlin DM, Miller CC (1996) The intracellular cytoplasmic domain of the Alzheimer's disease amyloid precursor protein interacts with phosphotyrosine-binding domain proteins in the yeast two hybrid system. Febs Lett 397:197–200

Mendelsohn AR, Brent R (1994) Applications of interaction traps/two hybrid systems to biotechnology research. Curr Opin Biotechnol 5:482–486

Meyer SC, Zuerbig S, Cunningham CC, Hartwig JH, Bissell T, Gardner K, Fox JE (1997) Identification of the region in actin-binding protein that binds to the cytoplasmic domain of glycoprotein IBalpha. J Biol Chem 272:2914–2919

Naik UP, Patel PM, Parise LV (1997) Identification of a novel calcium-binding protein that interacts with the integrin alphaIIb cytoplasmic domain. J Biol Chem 272:4651–4654

Osborne M, Dalton S, Kochan J (1995) The yeast tribrid system: genetic selection of trans-phosphorylated ITAM-SH2- interactions. BioTechnology 13:1474–1478

Ozenberger BA, Young KH (1995) Functional interaction of ligands and receptors of the hematopoietic superfamily in yeast [published erratum appears in Mol Endocrinol 1996 Aug;10(8):936]. Mol Endocrinol 9:1321–1329

Putz U, Skehel P, Kuhl D (1996) A tri-hybrid system for the analysis and detection of RNA–protein interactions. Nucleic Acids Res 24:4838–4840

Reymond R, Brent R (1995) p16 proteins from melanoma-prone families are deficient in binding to cdk4. Oncogene 11:1173–1178

Rothe M, Wong SC, Henzel WJ, Goeddel DV (1994) A novel family of putative signal transducers associated with the cytoplasmic domain of the 75 kDa tumor necrosis factor receptor. Cell 78:681–692

SenGupta DJ, Zhang B, Kraemer B, Pochart P, Fields S, Wickens M (1996) A three-hybrid system to detect RNA-protein interactions in vivo. Proc Natl Acad Sci USA 93:8496–8501

Shih HM, Goldman PS, DeMaggio AJ, Hollenberg SM, Goodman RH, Hoekstra MF (1996) A positive genetic selection for disrupting protein–protein interactions: identification of CREB mutations that prevent association with the coactivator CBP. Proc Natl Acad Sci USA 93:13896–13901

Song HY, Donner DB (1995) Association of a RING finger protein with the cytoplasmic domain of the human type-2 tumour necrosis factor receptor. Biochem J 309:825–829

Spencer DM, Belshaw PJ, Chen L, Ho SN, Randazzo F, Crabtree GR, Schreiber SL (1996) Functional analysis of Fas signaling in vivo using synthetic inducers of dimerization. Curr Biol 6:839–847

Spencer DM, Wandless TJ, Schreiber SL, Crabtree GR (1993) Controlling signal transduction with synthetic ligands. Science 262:1019–1024

Stanger B, Leder P, Lee T, Seed B (1995) RIP: a novel protein containing a death domain that interacts with Fas/APO-1 (CD95) in yeast and causes cell death. Cell 81:513–523

Stagliar I, Korostensky C, Johnsson N, te Heesen S (1998) A genetic system based on split-ubiquitin for the analysis of interactions between membrane protein in vivo. Proc Natl Acad Sci USA 95: 5187–5 192

Vidal M, Brachmann RK, Fattaey A, Harlow E, Boeke JD (1996) Reverse two hybrid and one-hybrid systems to detect dissociation of protein–protein and DNA–protein interactions. Proc Natl Acad Sci USA 93:10315–10320

Wang Z, Whitfield M, Ingledue T, Dominski Z, Marzluff W (1996) The protein that binds the 3′ end of histone mRNA: a novel RNA-binding protein required for histone pre-mRNA. Genes Dev 10: 3028–3040

Warbrick E (1997) Two's company three's a crowd: the yeast two hybrid system for mapping molecular interactions. Structure 5:13–17

Warbrick E, Lane DP, Glover DM, Cox LS (1995) A small peptide inhibitor of DNA replication defines the site of interaction between the cyclin-dependent kinase inhibitor p21WAF1 and proliferating cell nuclear antigen. Curr Biol 5:275–282

White MA (1996) The yeast two hybrid system: forward and reverse [comment]. Proc Natl Acad Sci USA 93:10001–10003

Xu C, Mendelsohn A, Brent R (1997) Cells that register logical relationships among proteins. Proc Natl Acad Sci USA 94:12473–12478

# Evolutionary Approaches to Protein Engineering

B. Steipe



Genzentrum der Ludwig-Maximilians-Universität, Feodor-Lynen-Str. 25, D-81377 Munich, Germany
e-mail: steipe@lmb.uni-muenchen.de

# 1 Targets and Tasks for Protein Engineering

The very term "protein engineering" remains something of an oxymoron, at least as far as engineering implies the rational application of well understood principles towards achieving a prespecified goal. Designed novel functions of proteins remain largely beyond our capabilities, despite intense efforts of numerous research groups in academia and industry. But we are trying, and the last few years have seen a rapid growth in the number of reports describing the successful application of a novel, and at the same time ancient, principle to the problem: evolutionary protein engineering. While this sophisticated trial-and-error approach may at first appear less rational than crystal structure gazing, it is undisputedly more successful – and there is nothing irrational about experimental success. This chapter will focus on the principles, concepts and methods of this field.

The targets for protein engineering have shifted over the last several years, away from medical applications and towards protein biotechnology, partly from concerns about the potential immunogenicity of engineered proteins, partly anticipating superior pharmacokinetic properties of small-molecule pharmacophores. Thus enzymes like proteases for washing powder formulations, amylases for food processing, cellulases and xylanases for pulp and paper processing are today's paradigms for industrial protein engineering (Rubingh 1997), while landmark applications in chemical synthesis (Moore et al. 1997), biosensors (Miesenböck et al. 1998) or bioremediation (Kumamaru et al. 1998) are appearing on the horizon. The situation may change, as protein engineering becomes an increasingly mature science and the first clinical trials of designed immunotoxins afford a view of the many things that are yet to come (Pai et al. 1996).

But what are the goals of engineering in the first place? A protein's role can be loosely divided into two aspects: folding and function, i.e. the intrinsic, structural aspects of the protein and its extrinsic interactions with its surroundings and both are targets for engineering (Table 1).

## 1.1 Folding

In general the prediction of the folded structure from sequence alone has been as elusive as the rational, targeted change of the sequence to produce novel structures or functions. It is only very recently that progress has been made. The protein folding problem results from the fact that proteins are context-sensitive complex systems, in which the precise effect of any sequence change is highly dependent on the interactions of the altered residue with its surroundings, including the solvent shell, and on the effects on the unfolded state. Predictions that would be based on a precise knowledge of these structures are accordingly difficult to make. Nevertheless, we can measure equilibrium and rate constants for the folding reaction and thus quantify the driving forces behind the phenomenon of self-organization of a polypeptide chain. Thus stability can be regarded as the metric of the protein folding problem.

**Table 1.** Examples of the diversity of properties that have been improved by evolutionary engineering

| Protein | Altered function | Reference |
|---|---|---|
| Barley α-amylase | Thermostability: ten-fold increase of half-life at 90°C | JOYET et al. 1992 |
| Subtilisin | Alkaline stability: doubling the autolytic half-time at pH 12 | CUNNINGHAM and WELLS (1987) |
| Subtilisin | Tolerates loss of stabilizing divalent cations | STRAUSBERG et al. (1995) |
| Subtilisin E | Active in 60% DMF | YOU and ARNOLD (1996) |
| *Streptomyces griseus* protease B | Broadened substrate specificity | SIDHU and BORGFORD (1996) |
| Green fluorescent protein | 40-fold brighter fluorescing bacterial colonies | CRAMERI et al. (1996) |
| Immunoglobulin constant domain | Preferential formation of heterodimers | ATWELL et al. (1997) |
| Immunoglobulin variable domain | Tolerates loss of structural disulfide bridge | MARTINEAU et al. (1998) |

### 1.1.1 Thermodynamic Stability

The concept of thermodynamic stability applies to an equilibrium between the native and the unfolded state. If, and only if, the folding reaction is completely reversible and satisfies the two-state approximation – no intermediate is more stable than either the folded or the unfolded state – we can simply count molecules in the folded state $F$ and the unfolded state $U$ and calculate the free energy difference $\Delta G$ between the two states (see, e.g. STEIPE et al. 1994 for an experimental protocol).

$$\Delta G = -RT \ln \frac{F}{U} \tag{1}$$

This is the purist's definition of stability. But for purposes of engineering we may be more interested in properties such as expression levels or inactivation rates and these will not have to correlate with thermodynamic stability in all cases.

### 1.1.2 Thermal and Environmental Stability

Thermostability is a desirable property in biotechnological applications for a number of reasons. Substrate solubility may be increased, the risk of microbial contamination may be minimized and the reaction rates may not only be increased in general, but may favor some side-reactions over others (COWAN 1997). Biotechnologically important processes may require extremes of pH, or the presence of chelators, proteases and detergents. Stability in aprotic environments would make protein catalysts interesting for a wide range of chemical transformations for which stereo- or regioselective catalysis is required. Unfortunately, most proteins denature only a few degrees above the physiological temperature and this is frequently an irreversible process that rapidly draws folded protein out of the equilibrium into the unfolded state. In general, the reason for irreversible inactivation at high temperatures or under other adverse environmental conditions is aggregation of the unfolded state. This process will be governed by the concentration and the unfolding rate $k_{\text{unfold}}$ which is itself related to equilibrium stability.

$$\Delta G = -RT \ln \frac{k_{\text{fold}}}{k_{\text{unfold}}} \tag{2}$$

Thus in any comparison of thermostability, care must be taken to consider the exact experimental conditions under which the measurements were performed. As illustrated in Fig. 1, mutations that lead to increased thermostability will affect the unfolding activation energy and need not necessarily increase the thermodynamic stability.

Solid experimental evidence exists today that the effect of point mutations on folding stability can be well approximated as additive, distributed, largely independent interactions. This situation is ideal for engineering, since the combination of stabilizing mutations can sum up to quite significant stabilization (ZHANG et al. 1995). Various methods have been described that allow the design of stabilizing mutations, such as the stabilization of α-helix macrodipoles (WALTER et al. 1995), the engineering of structural motifs like helix N-caps (AURORA and ROSE 1998) or β-turns (OHAGE et al. 1997), or the introduction of residues with higher intrinsic propensities for their respective conformational state (ZHANG et al. 1992), the introduction of disulfide bridges (JOHNSON et al. 1997), the reduction of the unfolded state entropy with X→Pro mutations (NICHOLSON et al. 1992) or the analysis of aligned sequence distributions (STEIPE et al. 1994). Engineering of protein stability is the one area in which rational engineering is competitive with evolutionary protocols.

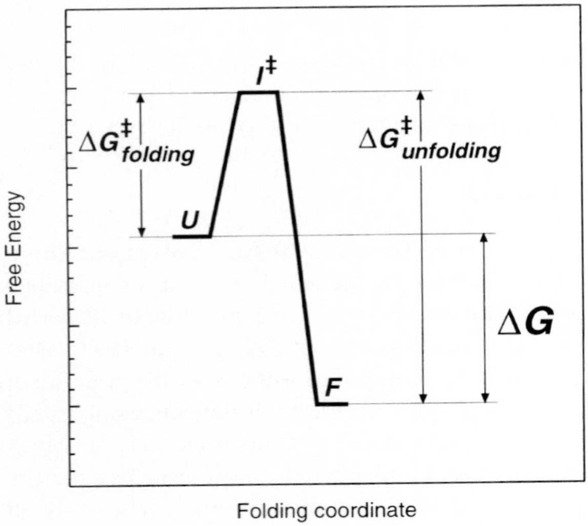


**Fig. 1.** Energy levels of the folding reaction. While the folding equilibrium is governed by the free energy difference, $\Delta G$, between the unfolded and the folded state, $U$ and $F$, the unfolding rate is determined by the free energy difference between the folded and the transition state, $\Delta G^{\ddagger}_{\text{unfolding}}$. Mutations can significantly affect either or all states, for instance a mutation that raises exclusively the transition state free energy will decrease the unfolding rate without changing the protein's thermodynamic stability

### 1.1.3 Other Folding Considerations

Properties other than thermodynamic or kinetic stability may need to be optimized for technological applications. The formation of structural disulfide bonds may be yield limiting, especially during in vitro refolding when free thiols may oxidize statistically, since non-native disulfide bonds will be favored. Unfortunately, the engineered removal of disulfide bonds carries a heavy energetic penalty, but with attention to refolding protocols very good yields can be obtained for many proteins (RUDOLPH and LILIE 1996). For proteins that posess a *cis*-peptidyl-prolyl bond in the native structure, in general the *trans-cis* isomerization during folding will be the rate-limiting step (SCHMID et al. 1996). Some of these *cis*-peptide bonds may be dispensable and engineering may be possible (KIEFHABER et al. 1990). Aggregation of intermediates during folding is the third major source of folding problems (KIEFHABER et al. 1991). Cellular chaperones have evolved to prevent aggregation in vivo, but engineering may also significantly improve the efficiency of folding. For example, the comparison of sequences of well-expressed immunoglobulin $V_H$ domains identified two residues that independently reduce domain aggregation in the periplasm and cell lysis during expression (KNAPPIK and PLÜCKTHUN 1995).

## 1.2 Function

Engineering a protein's function is significantly more challenging than stabilizing its structure. While stability can be improved with a number of independent, distributed point mutations – each of which may be a crude approximation to an optimal solution – the active site of a protein is typically localized to a unique discontinuous epitope, and function is exquisitely sensitive to the precise orientation and interactions of the participating residues. In this setting, mutations are no longer independent and their combined effects will deviate strongly from simple additivity. As a consequence, an evolutionary trajectory to a novel function may require crossing significant barriers of reduced activity. This has been demonstrated in an analysis of the catalytic triad of the serine protease subtilisin (CARTER and WELLS 1988). Both the substitution of the catalytic serine and histidine reduce the turnover number by a factor of approximately $10^6$; the combined mutations, as well as the substitution of the aspartic acid to alanine, have no additional deleterious effect. As a consequence, three coordinated amino acid changes would be needed to generate the catalytic triad and intermediates confer no selective advantage on the protein.

### 1.2.1 Binding

The most elementary interaction of a protein with its surroundings is binding another molecule. Highly complementary molecular surfaces have evolved to perform any biologically required task of binding and discrimination. How many epitopes will be required to bind any molecular shape with high affinity? This question bears directly on the design of evolutionary experiments. Experience with the immune system suggests that the number may be surprisingly small. While on the order of $10^7$ different combining sites may be generated in a primary immune response, as little as $10^2$–$10^4$ different B cells are sufficient to provide viral immunity (BACHMANN et al. 1994). Indeed, from phage display libraries with diversities of $>10^8$, domains with subnanomolar dissociation constants and off-rates of $10^{-3}$/s can apparently be isolated almost routinely (VAUGHAN et al. 1996; PINI et al. 1998).

### 1.2.2 Catalysis

To a significant degree, catalysis is just a different form of binding – binding the transition state of a chemical reaction and thus lowering its free energy (JENCKS 1969). This concept is well borne out by the catalytic activity of antibodies that have been raised against transition state analogs (LERNER et al. 1991). On the other hand, the catalytic function of antibodies has been consistently inferior to that of "true" enzymes that catalyze the same reaction. This is commonly believed to stem from the absence of functional groups that might take an active part in the reaction, but which do not increase affinity to the transition state analogue (WENTWORTH and JANDA 1998). Indeed, it can be shown that increased binding to the transition state analogue need not lead to increased catalytic rates (BACA et al. 1997). The lesson, not only for antibodies, is that successful evolutionary engineering of catalysts should involve direct selection for activity, and, in case this is difficult, more attention should be invested in the design of the selection process. Two strategies have been devised for catalytic antibodies that address this problem: reactive immunization (WIRSCHING et al. 1995; BARBAS et al. 1997) and mechanism-based panning of phage display libraries (Fig. 2) (JANDA et al. 1997).

In summary, function is governed by localized, highly cooperative interactions. Since even small differences in geometry may have dramatic effects on function, traditional structure-based engineering approaches have consistently failed to improve enzymes. For example, even something seemingly as straightforward as the reengineering of trypsin towards the substrate specificity of its close relative chymotrypsin has required major remodeling and transplantation of entire loops, comprising the substrate recognition subdomain (HEDSTROM 1996).

## 2 Concepts for Rational and Evolutionary Engineering Approaches

Protein engineering, whether rational or evolutionary, is the modification of an existing sequence for a new purpose. It is not trivial that protein engineering is possible at all.

a

b

c

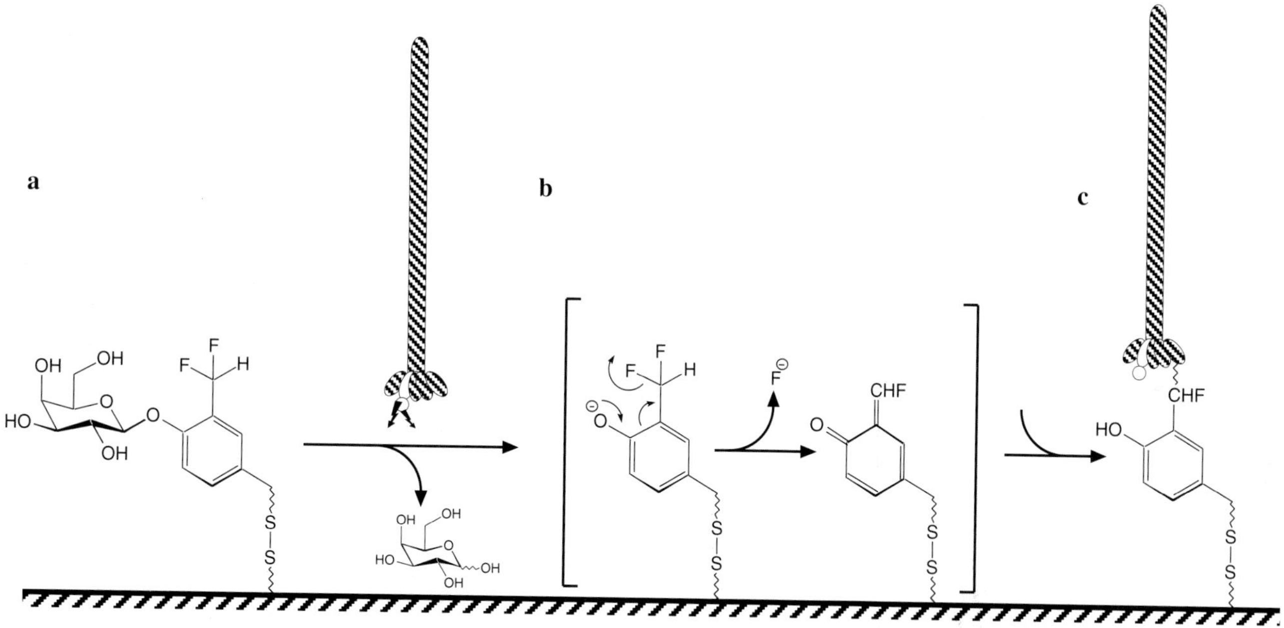

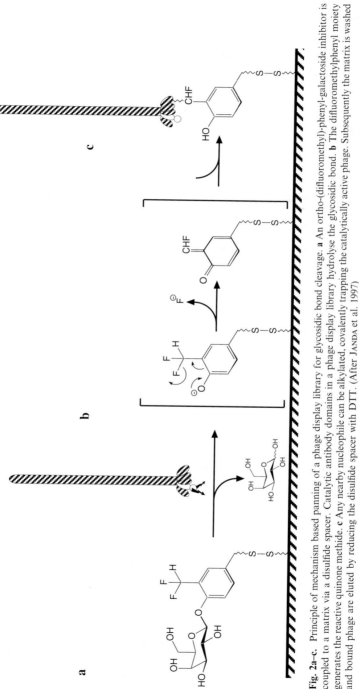

**Fig. 2a–c.** Principle of mechanism based panning of a phage display library for glycosidic bond cleavage. **a** An ortho-(difluoromethyl)-phenyl-galactoside inhibitor is coupled to a matrix via a disulfide spacer. Catalytic antibody domains in a phage display library hydrolyse the glycosidic bond. **b** The difluoromethylphenyl moiety generates the reactive quinone methide. **c** Any nearby nucleophile can be alkylated, covalently trapping the catalytically active phage. Subsequently the matrix is washed and bound phage are eluted by reducing the disulfide spacer with DTT. (After JANDA et al. 1997)

## 2.1 Theoretical Considerations

The basic assumption of protein engineering is that a natural sequence can be modified to improve a certain function. This implies: (1) that the protein is not already at an optimum for that function, otherwise it could not be improved; (2) that the required sequence changes can be accommodated without disrupting the structure, otherwise it would not fold; and (3) that the new sequence is not too different from the native sequence, otherwise it could not be found. None of these three observations is trivial. The first highlights the fact that evolution cannot generate proteins that are optimal for any given task, it can only generate proteins whose performance confers a selective advantage on the host organism. Thus optimization will cease when no more selective advantage can be gained from further improvement. In this view, proteins are not optimal but sufficient for their task and there is significant potential for improvement under guided selection. The second observation highlights the plasticity in protein structure and the redundancy inherent in a protein sequence. Most point mutations have only local, medium range effects and even though a large majority of mutations will be destabilizing, not all are completely disruptive. In fact there is a good reason that protein sequences should be optimized by evolution to be error-tolerant: the adaptability of a fold to random mutations is a factor determining the height of barriers on its evolutionary landscape. The third observation relates to the density of new optima in sequence space.

### 2.1.1 Sequence, Structure, Function Spaces and Landscapes

A sequence space is a very useful concept for the discussion of the evolution of proteins even though from a biochemist's perspective a sequence space has some very odd properties (see Kauffman 1993). Every sequence is considered to be represented by a point in space. The dimensions of the space correspond to the positions in the sequence, they at least equal the sequence length, but there may be more when gaps are needed. The dimensions are ordered, with an index increasing from the $NH_2$- to the COOH-terminal, and every dimension – the positions in the sequence – can take at least 20 different discrete values that have no obvious inherent ordering. Sequence spaces are large: a protein of 230 amino acids (233 amino acids is the median length of a protein chain in a library of 635 unrelated sequences in the structural database) spans a sequence space of $20^{230}$ or $10^{300}$ points. The relationship between two sequences can be described as their distance in sequence space; the size of a molecular library can be related to a volume surrounding its progenitor sequence.

Based on this concept, an evolutionary landscape can be defined by associating a function value, commonly called a fitness function, with every point in a sequence space. This can be a Boolean value, like "survival", a discrete value, like "number of oligomers", or, most frequently, a continuous value, like "melting point", "$k_{cat}$", or "racemic excess". Obviously, the function value and thus the shape of the landscape depends on the fitness function that is considered, or embodied in the experimental design. For example, on a landscape representing the catalytic rate towards the natural substrate, all the wild-type sequence neighbors in space are expected to perform worse than the wild-type, which is accordingly in a (local) sequence optimum. But the activity towards a different substrate implies a different fitness function and some direct sequence neighbors may improve on the wild-type in this case.

### 2.1.2 Evolutionary Trajectories

A walk in sequence space is a series of sequences, each derived from its predecessor in single steps or jumps. Such a walk can be random or adaptive, guided by increasing fitness function values. An evolutionary trajectory is the path between the initial and the final sequence. What elevation profile a trajectory will have will depend on the fitness function considered. The length of the trajectory is equal to the number of single point mutations in the evolutionary process.

As is evident from Table 2, the distance in evolutionary space that has been traversed in real, successful experiments is generally less than half a dozen steps. This can be interpreted in two ways: on the one hand, sequences for improved properties apparently lie close to their progenitors; they are not sparsely scattered in sequence space. On the other hand, the capacity of our present methods to generate functional sequences, more than a few point mutations away from a starting point, appears rather limited. Additionally, mutations are generated in a highly biased fashion.

Even if sequence space is to be sampled to a distance of only a few mutations, a strategy that relies on finding a successful variant by exhaustive search of a single, large sequence pool is likely to fail. The reason is a combinatorics problem: the codon dilemma. Amino acid sequence changes are encoded on a nucleotide level,

**Table 2.** Representative experiments using successive cycles of variation and selection

| Protein | Property | Number of cycles for success | Number of nucleotide changes required | Number of amino acid changes required | Reference |
|---|---|---|---|---|---|
| β-Lactamase | Increased activity | 3 | 4 | 4 | Stemmer (1994b) |
| GFP | Improved folding and expression | 3 | 3 | 3 | Crameri et al. (1996) |
| Subtilisin E | Stability in aqueous DMF | 2 | 3 | 3 | You and Arnold (1996) |
| Arsenite membrane pump | Increased activity | 3 | 3 | 3 | Crameri et al. (1997) |
| FLP-recombinase | Thermostability | 8 | 3–4 | 3–4 | Buchholz et al. (1998) |

In almost all cases a single nucleotide change leading to a single amino acid change was sufficient per cycle, the number of silent mutations was approximately the same. No amino acid change was reported that would have required more than one nucleotide change. Thus current protocols appear to sample sequence space in a biased fashion, in single mutation steps.

and a single amino acid change may require up to three coordinated changes of the coding sequence. The consequences can be dramatic: from the perspective of the amino acid sequence, the probability for a specific change in a sequence of length 230 is:

$$p = \frac{1}{\substack{\text{sequence}\\ \text{length}}} \cdot \frac{1}{\substack{\text{amino acid}\\ \text{alternatives}}} = \frac{1}{230} \cdot \frac{1}{19} = 2.3 \cdot 10^{-4} \tag{3}$$

But from the genetic perspective, the average probability for encoding a specific amino acid change through random nucleotide changes depends strongly on the number of required nucleotide changes:

$$p \approx \left( \frac{1}{\substack{\text{gene}\\ \text{length}}} \cdot \frac{1}{\substack{\text{nucleotide}\\ \text{alternatives}}} \right)^{\binom{\text{number of}}{\text{changes}}} = \left( \frac{1}{230 \cdot 3} \cdot \frac{1}{3} \right)^{d}. \tag{4}$$

$d = 1$ in 40% of mutations, e.g. Tyr(TAC) $\rightarrow$ Phe(TTC);

$$p = \left( \frac{1}{230 \cdot 3} \cdot \frac{1}{3} \right) = 4.8 \cdot 10^{-4}.$$

$d = 2$ in 53% of mutations, e.g. Tyr(TAC) $\rightarrow$ Trp(TGG);

$$p \approx \left( \frac{1}{230 \cdot 3} \cdot \frac{1}{3} \right)^{2} = 2.3 \cdot 10^{-7}.$$

$d = 3$ in 7% of mutations, e.g. Tyr(TAC) $\rightarrow$ Met(ATG);

$$p \approx \left( \frac{1}{230 \cdot 3} \cdot \frac{1}{3} \right)^{3} = 1.1 \cdot 10^{-10}.$$

Surprisingly, even a large library by common laboratory standards, say $10^8$ sequences, will not exhaustively encode all single point mutations! The consequences are: since one can only expect to densely sample sequence space to a distance of one, at best two, mutations, the successful application of evolutionary engineering requires that an evolutionary path exists that will yield a detectably improved function for every single evolutionary step.

Indeed, natural evolution works fundamentally along the principle of achieving results against impossibly small odds by arriving at the target sequence in stepwise improvements. The true power of evolutionary engineering lies in devising methods to iterate variation and selection.

If the improved function requires three or more cooperatively interacting sequence changes – every individual mutation being deleterious – then the chances of traversing such a barrier become vanishingly small. When this must be suspected, efforts should be focused on reducing the volume of sequence space that is to be searched, e.g. by developing some hypothesis on which region of the protein should be targeted, or by employing some scheme of site-directed random mutagenesis.

### 2.1.3 Search in Sequence Space

A model well suited for a theoretical investigation of the structure of molecular fitness landscapes and search trajectories was introduced by S. Kauffman (KAUFFMAN 1993). His *NK* model considers sequence spaces for sequences of a length of *N* sites. Each site can take *A* states and makes a contribution to the overall fitness of the sequence that depends on its own state and that of *K* other sites. When $K=0$, the sites contribute independently and additively to the global fitness, when *K* is maximal, i.e. $K=N-1$, each site is influenced by every site. While the value for *K* in natural proteins or even peptides is different for every site and has not been well determined experimentally, computer models that vary *K* can shed some light on the ruggedness of the evolutionary landscape and suggest efficient ways to locate minima. In a comparison of pooling, recombination and mutation strategies for an *NK* model of a random hexapeptide library, the available experimental data apparently support a value of *K* around $0.5 \times N$ – intermediate between being random and fully correlated (KAUFFMAN and MACREADY 1995). The landscape for $K=0$ is smooth, possessing a single peak which can be readily found. For small *K*, sequences in a local optimum will be fitter than most one- or two-mutant neighbors. The larger *K* is, the more likely it is that an evolutionary trajectory will become trapped in a local optimum and the probability for finding improved sequences becomes independent of search distance – the landscape is then uncorrelated. Conversely, for small *K*, i.e. correlated landscapes, the probability of finding an improved sequence decreases with search distance. How does this translate into the vocabulary of molecular biology? Properties that require a significant number of cooperative interactions before an improvement in fitness is observed cannot be found by any strategy that is currently practical. Properties that can be improved with independent or quasi-independent point mutations have a good chance to be selectable in iterated evolutionary cycles. In this case, single or double mutations per cycle search sequence space more efficiently than more radical changes.

The most important conclusion is the importance of investing more effort in the design of the experimental protocol: being able to detect even slight advantages in the desired function and running the evolutionary optimization through a large number of cycles, rather than constructing ever larger libraries.

## 2.2 Complementing Evolutionary Approaches with Rational Concepts

Rational engineering designs solutions top down: it is an attempt to divine the location of the desired optimum and to design experiments according to this insight. Rational engineering requires knowledge of the sequence and preferably the structure of the protein, delineation of the active site, understanding of the mechanism, identification of cofactors, etc. Most importantly, it requires a hypothesis about the limiting step for the desired function. In well characterized systems, the performance of rational engineering can be quite remarkable. As an example, the thermolysin-like protease (TLP) has been engineered with eight point mutations for thermostability to resist boiling temperatures (Van den Burg et al. 1998). Remarkably, the mutant enzyme is as active at room temperature as the wild-type. Individual mutations were contributed from an analysis of sequence differences to thermolysin, from an increase in the number of residues that lower the entropy of the unfolded state and from a designed disulfide bridge – tried and proven approaches to rational protein engineering.

State of the art protein engineering and design applies some computational algorithm, an objective function, to a novel sequence and then attempts to find an improved sequence through methods of combinatorial optimization. That this process is becoming practical is evident from the successful de novo design of a protein G-β1 domain that is 18kJ/mol more stable than the wild-type (Malakauskas and Mayo 1998). Thus modern protein design in its application of combinatorial optimization principles frequently is itself in silico evolutionary engineering.

Two main benefits of rational design for evolutionary engineering can be identified: the first is the possibility to construct stable structural frameworks for the display of combinatorial libraries, the second is the generation of hypotheses that allow limiting the required size of the library, such as constraining diversity to spatially adjacent residues or conserving hydrophobicity profiles.

## 3 Evolutionary Engineering Methods

Evolution implies iteration, and the practical application of evolutionary principles to protein engineering involves repeating cycles that can be divided into three parts: the generation of genetic diversity, the coupling of genotype and phenotype and the identification of successful variants.

### 3.1 Generating Diversity

The exhaustive mutation of a limited number of sites is a fundamentally different experiment from the stepwise optimization of entire genes. The former case can avoid the codon dilemma: sequence space can well be sampled exhaustively to five or six positions. The disadvantage is that only a subset of the entire gene can be targeted. For this reason, degenerate oligonucleotides are commonly used in the construction of epitope libraries, while diversity in libraries of entire proteins is commonly generated with some PCR-based procedure.

#### 3.1.1 Oligonucleotide Directed Mutagenesis: Circumventing Genetic Code Degeneracy

If only short regions of the protein are to be targeted, various methods of directed mutagenesis with degenerate oligonucleotides can be employed. The simplest and oldest approach is to use equimolar mixtures of all four nucleotides, (N)(N)(N), for the codons that are to be changed (OLIPHANT et al. 1986), but this may not be the best strategy. In procedures that involve the synthesis of a complementary strand, a bias for incorporating the original nucleotide will arise from the preferential hybridization of oligonucleotides that form larger numbers of Watson-Crick base pairs. This bias can be eliminated by reducing the concentration of the wild-type nucleotide during synthesis (AIRAKSINEN and HOVI 1998). But more importantly, an (N)(N)(N) codon mixture is biased in favor of those amino acids with more entries in the genetic code table; for instance, it will contain six times more leucine than methionine and it will contain 4.7% stop codons in every position. Thus, the chance of arriving at a randomized sequence of length $N$ that can be translated without stop codons is:

$$p = \left(1 - \frac{3}{64}\right)^N \tag{5}$$

e.g. $p \approx 0.6$ for ten residues and $p \approx 0.4$ for 20 residues. If the library is large enough to contain every variant sequence and the selection process can pick out individual sequences, these shortcomings will not be relevant. But if the library can sample sequence space only sparsely, more intelligent strategies are needed to improve its diversity and quality. A useful alternative is the codon mixture (N)(N)(C,G,T). This mixture not only encodes a more even distribution of amino acids, but also reduces stop codon frequency to 2% – improving the chances for a translatable sequence to $p \approx 0.8$ for ten residues and $p \approx 0.67$ for 20 residues. Alternate schemes have been published that exploit the error-tolerance features inherent in the genetic code. Mixtures can be biased towards residues with common physicochemical properties such as size, hydrophobicity or charge while at the same time the redundancy is reduced (BALINT and LARRICK 1993) (Table 3). Biasing amino acid distributions requires a hypothesis about which choices are advantageous. This may be based on sequence alignments of homologous genes, on conserving the physicochemical properties of the mutated residues (e.g., Table 4), or, as in a procedure termed "recursive ensemble mutagenesis" (DELAGRAVE et al. 1993), on compiling the distributions from the sequence pool of the preceding evolutionary cycle.

**Table 3.** A codon mixture for charged amino acids (BALINT and LARRICK 1993)

| Charged (R)(R)(K) | | | |
|---|---|---|---|
| Position | 1 | 2 | 3 |
| A | 50% | 50% | 50% |
| C | – | – | – |
| G | 50% | 50% | – |
| T | – | – | 50% |

Resulting amino acid spectrum (probability): acidic, E(0.125) D(0.125); basic, R(0.125) K(0.125) H(–); hydrophilic, Q(–) N(0.125) T(–) S(0.125); hydrophobic, V(–) L(–) M(–) I(–) Y(–) W(–) F(–); small, A(–) G(0.250); problems, P(–) C(–) Stop(–).

Note that all encoded amino acids except glycine are present in the mixture with equal probabilities. Stop codons are excluded. If a charged residue is required with certainty at the targeted position, this mixture is nearly optimal, as far as simplicity of synthesis and complexity is concerned.

**Table 4.** A codon mixture centered on glutamate for the construction of evolutionary libraries

| Glu (at $d=0.5$ of average distance in BLOSUM 62 matrix) | | | |
|---|---|---|---|
| Position | 1 | 2 | 3 |
| A | 38% | 41% | – |
| C | 27% | 18% | 30% |
| G | 31% | 21% | 53% |
| T | 4% | 20% | 17% |

Resulting amino acid spectrum (probability): acidic, E(0.291) D(0.137); basic, R(0.028) K(0.112) H(0.048); hydrophilic, Q(0.102) N(0.053) T(0.029) S(0.007); hydrophobic, V(0.023) L(0.008) M(0.006) I(0.003) Y(0.002) W(0.0005) F(0.0001); small, A(0.074) G(0.046); problems, P(0.026) C(0.0003) Stop(0.005).

Note the good correspondence of the amino acid frequencies with physicochemical measures of similarity – charge, hydrophobicity and volume. The frequency for nonsense mutations is reduced by a factor of ten relative to its occurrence in a random nucleotide mixture. No amino acid is completely excluded. The mixtures are adjusted to compensate for unequal reactivity of nucleotides during synthesis. Such mixtures can be optimized individually for every amino acid and synthesized on standard oligonucleotide synthesizers that allow independent control of reagent concentrations.

Under specific circumstances it may be desirable not to approximate amino acid distributions with degenerate codons, but to specify them explicitly by synthesizing random libraries directly from building blocks of trinucleotides. That such an approach is indeed feasible, after careful optimization of the synthesis strategy, has now been reported by a number of groups (VIRNEKÄS et al. 1994; LYTTLE et al. 1995; ONO et al. 1995; KAYUSHIN et al. 1996; GAYTAN et al. 1998). The downsides of this method are that it is not commercially available and it does not solve the problem of deciding which mixture of amino acids may be desirable at any given position.

In order to combine ease of synthesis with balanced mixtures of amino acids, redundancies and symmetries in the genetic code can be favorably exploited (Steipe and Bruhn, in preparation). A typical application would be the synthesis of conformationally constrained epitopes, such as antibody combining sites, for which some preference for each position can be defined, e.g. conserving the hydrophobicity profile, yet no amino acid should be rigorously excluded, since it might be just the one critical for the desired function. In effect, amino acid properties such as hydrophobicity or size should be statistically constrained to limit the destabilizing effect of the new sequence on the framework structure. Thus similar residues should be more frequent in the mixture than dissimilar residues. To achieve this, nucleotide mixtures can be simultaneously optimized in every codon position, to maximize the resulting codons' complexity – defined as the information-theoretical information content – and to limit their dissimilarity, measured empirically in terms of exchange probabilities from a mutation data matrix (Table 4).

### 3.1.2 Chemical Mutagenesis, Mutator Strains and UV Irradiation

While these methods were among the earliest used in evolutionary engineering (SINGER and KUSMIEREK 1982), they have been largely superseded by the more modern techniques described below. The main disadvantage of all three methods is their indiscriminate targeting of the entire genome (or at least an entire plasmid), and this makes it rather likely that the screen for function will be influenced by non-specific effects, like altered expression rates, or even the spontaneous modification of cellular enzymes to perform the task that is being screened or selected for.

### 3.1.3 Error-Prone PCR

For most purposes, the introduction of nucleotide changes via error-prone PCR will be the method of choice: it is simple, efficient, restricted to the region of interest and well characterized. The protocol devised by LEUNG et al. (1989) was subsequently improved (CADWELL and JOYCE 1994) to reduce the inherent bias of nucleotide transitions over transversions (SHAFIKHANI et al. 1997). Both methods allow tuning the mutation rate by varying the concentrations of $Mn^{2+}$ and dNTPs, or the number of PCR cycles. Since beneficial mutations are rare and the combination with a disruptive mutation will produce an inactive protein, it is best to keep the mutation rate to a level of one or two sequence changes per gene. Since some amino acid changes will require three concerted nucleotide substitutions, in practice the ideal number of nucleotide changes is between two and six over the length of the gene. Very high mutation rates for the randomization of short epitopes can be achieved with the inclusion of synthetic nucleoside analogues, that can base-pair ambiguously (ZACCOLO and GHERARDI 1996).

### 3.1.4 DNA Shuffling

By far the most successful approach to molecular evolution appears to be the DNA shuffling method, pioneered by W.P. Stemmer to address the question of how an efficient walk on an evolutionary landscape can be generated experimentally (STEMMER 1994a) (Fig. 3). This protocol allows successful mutations to be passed among sequences by recombination. The power of this "sexual PCR" process stems from the possibility of preserving locally optimal solutions, which may improve the desired property synergistically when combined.

In a first step, the gene of interest is cleaved into many short, random fragments with DNAse I. These fragments of 10–50 base pairs are then purified and recombined in a PCR-like process without exogenous primers. Terminal primers are added to the last step of extension and full length sequences are amplified and cloned. Since the melting and annealing steps will cause fragments from different strands to hybridize, an efficient recombination of strands takes place. In principle, this process would simply regenerate the native sequence, but variation can be introduced into the pool by various processes:

1. By initially amplifying the wild-type gene before fragmentation under mutagenic PCR conditions
2. By initially using a pool of genes, such as homologous genes from different organisms (CRAMERI et al. 1998)
3. Intrinsically, by the process of extension and recombination itself, which has an intrinsic, tunable error-rate (ZHAO and ARNOLD 1997)

4. By the addition of mutagenic primers to the mixture, and
5. By pooling the evolved genes from the most successful mutants of each evolutionary cycle

After the successful isolation of mutants, further cycles of recombination can be performed with an excess of the wild-type sequence under stringent selection conditions. This process of back-crossing will revert nonessential mutations to the wild-type sequence and thus give some insight into the essential sequence changes.

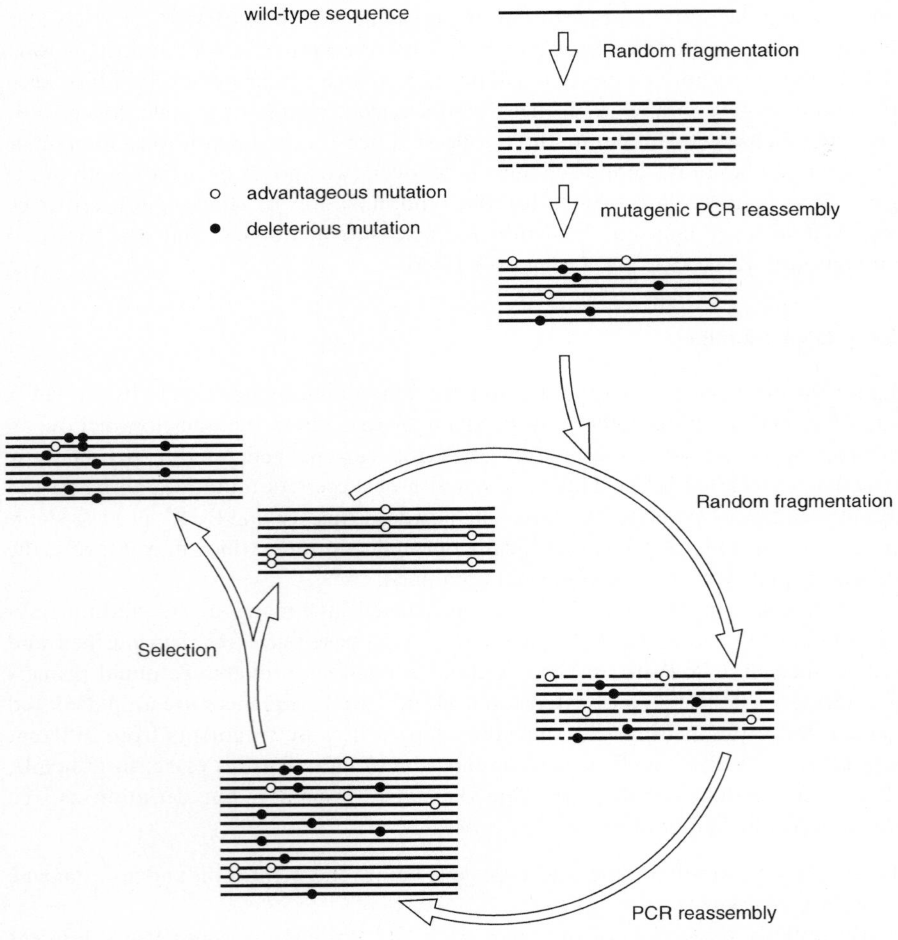


**Fig. 3.** The DNA shuffling method for molecular evolution. After random fragmentation, a pool of genes is reassembled with a PCR protocol that at the same time is mutagenic and generates multiple recombination events. From the recombined library, functional sequences are selected and the new resulting pool is improved by further iterations of the protocol

### 3.1.5 Recombination In Vivo and In Vitro

DNA shuffling is modeled along natural recombination, but can natural recombination itself be used for the generation of large molecular libraries? Three site-specific recombination systems have been analyzed in some detail in *E. coli*: phage lambda Int, transposon Tn3 and the Cre recombinase of bacteriophage P1. Of these, the Cre-loxP system appears ideally suited for engineering purposes (Hoess et al. 1984). It is simple, requiring only 34 bp of DNA binding site and the recombinase, and it appears to work independently of local DNA conformation.

In a particularly elegant application, Fisch et al. (1996) reported the generation of a large, combinatorial peptide library. Two artificial exons, each encoding ten randomized amino acids, were joined via a five residue spacer and fused to the pIII phage coat protein for phage display. Exon shuffling was achieved via the lox recombination site cloned into a self-splicing group I intron (Cech 1990), which automatically excises itself after transcription. The authors report a library size in excess of $10^{11}$ peptides and note the potential of the system for the de novo evolution of small peptides and proteins.

Recombining a set of highly homologous genes in vitro can also be achieved in a procedure called staggered extension process (StEP) recombination (Zhao et al. 1998). StEP involves a PCR procedure with a low concentration of terminal primers or random-sequence primers (Shao et al. 1998) and very short extension cycles at reduced temperature, which will only extend primers over 5–20 base pairs per cycle. These abbreviated fragments will switch templates during the denaturation/annealing cycles and the final, full-length sequence will have been synthesized from a number of different templates.

Whether in vitro recombination is achieved by template switching or by DNA shuffling, its capacity to accelerate the search process makes it the core of modern, efficient evolutionary protocols. Initial point mutations sample local new optima of the evolutionary landscape. Successful variants can subsequently be combined, removing silent and deleterious mutations and further increasing activity (Moore et al. 1997). The combined mutations put the protein into a more distant region of sequence space, one that would not previously have been accessible with a library of practical size. At this new optimum, the process can be repeated until the limiting factor is the sensitivity of the experimental setup to identify further improvement.

## 3.2 Coupling Genotype and Phenotype

In order to identify desired sequences, some strategy needs to be devised that will ensure that the desired function will be in some way physically associated with its gene. The alternative of direct sequencing of the improved protein is currently not technically feasible, even though it has been successfully applied to the analysis of peptide libraries on beads (Lam et al. 1991). Similarily, encoding schemes have been developed for non-genetic combinatorial libraries (Czarnik 1997). Yet another similar concept is embodied in the synthesis of peptide libraries on beads, together with a synthetic oligonucleotide encoding the sequence (NEEDELS et al. 1993). The advantage, as in all procedures based on combinatorial chemistry, is the possibility to incorporate non-proteinogenic amino acids; but the chief disadvantage is the limited library size and the added difficulty of decoding as compared to genetically based methods. An array of methods to couple information and function are described below, ranging from the binding of expressed peptides to their genes to the association of gene and protein in living cells.

### 3.2.1 RNA-Peptide Fusions

Covalent fusions of an mRNA and its encoded peptide can be achieved when a pool of mRNAs is synthesized with the peptidyl-acceptor antibiotic puromycin attached to the 3′ end. The mRNAs are in vitro translated, the 3′ puromycin end – an analog to a charged tRNA – will bind to the ribosomal A site at some time during the translation and its free amino group will be transferred to the carboxylate end of the nascent peptide chain (NEMOTO et al. 1997). Once this has happened, the mRNA is covalently bound to the peptide and the adduct will dissociate from the ribosome. These adducts can then be screened for the desired function, the mRNA of successful sequences reverse-transcribed, amplified and cloned for analysis (ROBERTS and SZOSTAK 1997). This procedure has tremendous potential, since library sizes of $10^{12}$ have been achieved and $10^{15}$ should be attainable with some optimization and scale-up; they are thus far larger than those obtainable with other methods. We are certain to see reports of refinements soon, such as the use of longer sequences, or even obviating the need for in vitro mRNA-puromycin synthesis, perhaps through the use of a ribozyme sequence. The only downside appears to be, in principle, the requirement for single molecule detection efficiency.

### 3.2.2 Ribosome Display

An alternative to the chemical coupling of mRNA and peptide is to preserve their association on the ribosome. This procedure has been developed for peptide libraries (MATTHEAKIS et al. 1996) and for functional proteins (HANES and PLÜCKTHUN 1997; HE and TAUSSIG 1997). While the procedure requires some biochemical sophistication, the large library sizes of $>10^{12}$ individual molecules and the possibility to use full-length proteins makes it very attractive. No additional transformation steps are required and PCR amplification between cycles of enrichment allows the introduction of random mutations – evolutionary engineering, entirely in vitro.

### 3.2.3 Peptide on Plasmid

One of the simplest in vivo embodiments of the coupling of information and structure is the direct, physical association of the target molecule with its gene via a DNA-binding domain. For peptide libraries, this has been achieved with the fusion of a library to the C-terminus of the *lac*-repressor (CULL et al. 1992; SCHATZ et al.

1996). After isolation of the repressor-plasmid complex from the cell, ligand binding candidates can be retained on an affinity column. The plasmid can be eluted either by denaturation or by adding the inducer and the eluate used to transform cells. Note that even though the number of molecules participating in the experiment can be large, the actual diversity of the library will equal the number of cells into which the initial library has been transformed.

While the procedure is conceptually extremely simple and powerful, requiring no additional decoding or cloning steps, it requires a gentle, yet quantitative procedure of lysing cells that will not interfere with tight physical association of the protein with the DNA, and it requires a slow off-rate of the DNA-binding domain to prevent exchange of the binders. The procedure is also sensitive to interference from intrinsic DNA-binding properties of the target or the matrix.

Another potential problem arises from the fact that LacI dimerizes via its 300-amino acid COOH-terminal domain. Dimerization is a disadvantage for screening, because avidity effects can result in the selection of intermediate- to low-affinity interactors which outnumber high-affinity binders. Thus a monomer domain would be desirable. Such a protein has been constructed by evolutionary engineering of a synthetic linker peptide that fuses two 60-amino acid DNA binding 'headpiece' domains of LacI. This monomeric protein binds DNA stably and can be used for panning and enrichment of high-affinity binding peptides (GATES et al. 1996).

### 3.2.4 Phage Display

The most widely used system for screening libraries today is phage display (BURTON 1995). Peptides or protein domains are fused (most commonly) to the gene III protein (gIIIp) of filamentous phage. After the host cell is infected by helper phage, the fusion protein is incorporated into newly made phages together with its coding gene. Functional sequences are subsequently enriched from the pool by binding a ligand matrix and eluted phages can be directly transformed into host cells for amplification and analysis. The system is very versatile: besides peptide libraries, successful fusions have been reported for the engineering of enzymes such as alkaline phosphatase (MCCAFFERTY et al. 1991), β-lactamase (SOUMILLION et al. 1994), staphylococcal nuclease (LIGHT and LERNER 1995) or even trypsin (WANG et al. 1996). Protease inhibitors have been engineered (MARKLAND et al. 1996) as successfully as cytokines (VISPO et al. 1997), growth hormone (CHIEN et al. 1991), and zinc-finger domains (REBAR and PABO 1994). A particularly interesting new development is the recruitment of a lipocalin framework for the display of a large epitope library (BESTE et al. 1999). These novel proteins have been aptly called anticalins by the authors. The largest area of application, however, has been immunoglobulin domains, particularly single-chain Fv fragments (scFvs) (see HOOGENBOOM et al. 1998 and GRIFFITHS and DUNCAN 1998 for recent reviews).

In general, stable cytoplasmic proteins appear to cause problems in this system. The assembly of filamentous phage takes place in specific assembly sites where inner and outer membranes come in close contact and the proteins involved are stored as integral proteins of the inner membrane until they are incorporated into the growing phage. Both overexpression of the fusion protein as well as fusions with proteins that cannot unfold for secretion will lead to toxicity and ultimately to plasmid instability, degrading library diversity or leading to complete loss of the molecule to be displayed. Two factors have been identified that appear to alleviate the problem. First, it is important to repress the background expression as completely as possible, before induction of phage assembly. For example, this can be achieved through introduction of a transcriptional terminator upstream of the *lac* promoter (KREBBER et al. 1996); other tightly regulated promoters (SKERRA 1994) may serve the same purpose. Second, efficient secretion of the fused protein appears to be crucial for efficient display and this includes translocation across the inner membrane as well as folding in the periplasmic space. By panning a library of *E. coli* proteins, coexpressed with a poorly folding scFv-gIIIp fusion, for high expression levels, BOTHMAN and PLÜCKTHUN (1998) were able to identify a protein that improves the expression of a wide range of scFv fragments by increasing the amount of displayed protein. Significantly, the protein indeed is a periplasmic chaperone, the *skp* or *ompH* gene product (CHEN and HENNING 1996), reminiscent of earlier reports that coexpression of the cytoplasmic chaperone GroE would increase phage titers by two orders of magnitude (SODERLIND et al. 1993).

Fusions to gIIIp have the advantage of monovalency, since gIIIp is present in only five copies and the native gIIIp is supplied in excess by the helper phage, but this may not be desirable in all cases. Alternatives for multivalent display have been described, such as fusions to gene VIII protein, the major coat protein of filamentous phage (MAKOWSKI 1994), or fusions to the D protein of the phage lambda capsid (STERNBERG and HOESS 1995).

Selectively infectious phages (SIP) have been constructed, which obviate the panning step (SPADA et al. 1997). The $NH_2$-terminal domain of gIIIp is replaced by the protein library, e.g. an scFv, while the ligand is chemically coupled or genetically fused to $NH_2$-terminal domains (Fig. 4). While the phage itself is non-infectious, the interaction of a protein from the library with the ligand restores infectivity. Based on a similar concept, ligand epitopes have been expressed as fusions to the tip of the bacterial F pilus. While this abolishes infectivity of wild-type phage, phages displaying an scFv against the peptide epitope became selectively infectious (MALMBORG et al. 1997).

### 3.2.5 Cell-Surface Display

Many thousands of copies of protein or peptide libraries can be displayed on the surface of cells. Thus such libraries can be targeted with a fluorescent labeled ligand, the cells sorted by FACS, and grown, obviating amplification or transformation steps (FUCHS et al. 1996). In contrast to phage display, which may only recover less than $10^{-3}$ of library elements, cell-based systems can almost guarantee quantitative recovery of library elements. With the variety of proteins now available that can accept extensions or insertions, cell-based screens can be predicted to rapidly grow in importance (GEORGIOU et al. 1997).

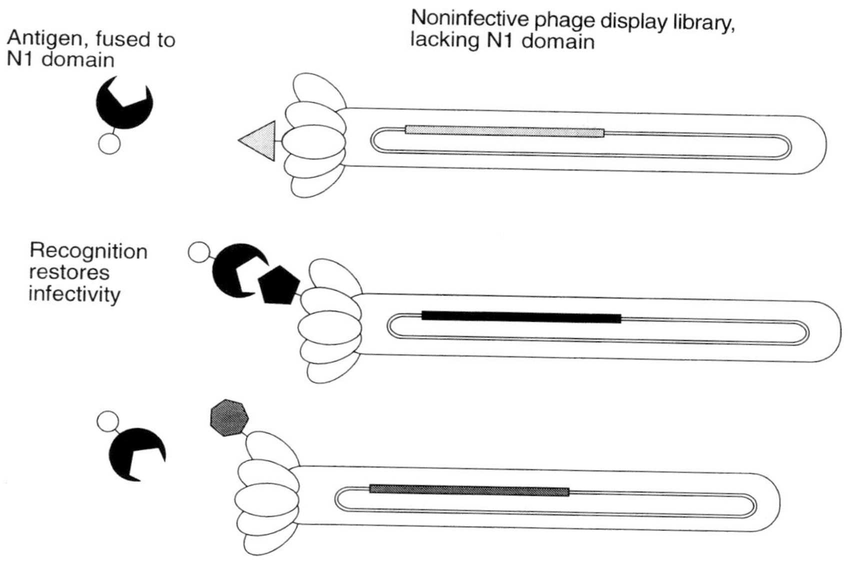


**Fig. 4.** Selectively infectious phage display

### 3.2.6 Micro-compartmentalization

It may be attractive to abstract the principle of micro-compartmentalization from cells to artificial systems. Beyond coupling information and function, entire pathways may be spatially isolated. For instance, using a simple spray-gun, droplets of 50–200nl volume can be generated that may contain substrates, cells and even synthesis beads (Borchardt et al. 1997). Stable preparations can be achieved by encapsulating the desired molecules and reactions in liposomes or in oil–water emulsions. Liposomes are the closest artificial models of cells and may even withstand the elevated temperatures required for PCR (Oberholzer et al. 1995), thus they may have potential for allowing multiple evolutionary cycles to be performed in situ. But water-in-oil emulsions may be even simpler to prepare and their use for molecular evolution has already been demonstrated by an in vitro enrichment of DNA methyltransferase genes from a $10^7$-fold excess of DHFR genes (Tawfik and Griffiths 1998). Under the conditions reported by the authors, the mean droplet diameter was 2.6μm – on the order of a bacterial cell – and there were approximately $10^{10}$ compartments formed per milliliter.

## 3.3 Screening and Selection

Screening is the identification of active variants by comparing them with all other elements in a molecular library. Selection is the enrichment of active variants in a molecular library. Since screening, in principle, requires assaying every single individual, the size of screenable libraries will be limited in practice to $10^5$–$10^7$ sequences. Selection procedures may be used on much larger libraries and they may be more sensitive than screens, frequently requiring less than 1% of background activity for success. However, they require that the desired activity can be linked somehow to a significant growth advantage. Finally, selective pressure on living cells will induce a variety of responses, and the appearance of random phenotypic variants displaying the desired property must not be significantly more probable then its emergence as the result of library evolution.

### 3.3.1 Screening

Since the size of screenable libraries is limited (i.e. one can grow on the order of $10^5$ isolated colonies on a large petri dish), to screen large libraries either the evolutionary process has to be broken down into more cycles, successful variants have to be enriched by a preselection step or individuals have to be pooled and singled out in successive steps.

Screening commonly relies on visual detection and much ingenuity has gone into the design of protocols that couple some function to a visual signal, commonly via activation of a reporter gene. The three most commonly used reporter enzymes are β-galactosidase, chloramphenicol acetyl transferase and luciferase (GROSKREUTZ and SCHENBORN 1997). Various substrates are available for chromogenic enzymatic reactions; they are most frequently based on color changes of a nitrophenol leaving group which is released by hydrolysis of a substrate, or the precipitation of an insoluble, blue indigo dye (e.g. X-gal or BCIP/NBT). Protease activity can frequently be directly visualized by the formation of halos around colonies grown on casein- or skim milk-agar. This principle has been used successfully to screen variants of subtilisin E with increased activity (YOU and ARNOLD 1996).

Fluorescence-based screening methods are rapidly gaining importance, since they provide very high sensitivity, down to single-molecule detection, together with low background (EIGEN and RIGLER 1994). Fluorophore binding proteins are a common model system for molecular evolution since affinities can be well determined (HENNECKE et al. 1998; BESTE et al. 1999). Fluorogenic enzyme substrates have been in use for a while, with 4-methylumbelliferone being a common fluorophore. An interesting new development is the use of intramolecular fluorescent resonant energy transfer (FRET) for the detection of catalysis (ZLOKARNIK et al. 1998). In this work, a β-lactam-based fluorogenic substrate with a large emission wavelength shift after hydrolysis was synthesized. The expression of as few as 100 β-lactamase molecules per single cell can be detected, making this an extremely sensitive and versatile system to monitor gene expression.

Undoubtedly the most important contribution to fluorescence based screening has come from green fluorescent protein (GFP) (TSIEN 1998). In only 4 years, this protein has become a standard component of the tool kits of cell biologists and protein engineers alike, and the ready visual identification of variants has made it one of the important models of evolutionary engineering. For example, an error-prone PCR amplification will produce the mutation Tyr66His with a frequency of

$\approx 10^{-4}$, a blue fluorescent mutant (BFP), which can be well distinguished from the wild-type with a hand-held UV-lamp. Wavelength shifted mutants of GFP can be used as the basis for genetically expressible intracellular sensors. For example Roger Tsien's group has fused GFP and BFP to calmodulin, constructing a sensitive FRET-based calcium sensor (Miyawaki et al. 1997), while Miesenböck et al. (1998) have used evolutionary methods and microtiter plate screens to evolve intracellular pH sensors.

Bioluminescence assays complement fluorescence as reporter systems for gene expression. Their substrates can be synthesized by the host after supplying the necessary genes in *trans* on a separate plasmid (Manen et al. 1997), and a wide variety of commercially available cloning vectors exist (Groskreutz and Schenborn 1997).

Screening has traditionally been an analysis of single bacterial colonies. The colonies can either be directly visualized, or lysed, blotted and a variety of immunochemical methods applied. A particularly elegant example utilizing colony blots, is the optimization of streptavidin to bind a peptide tag, by randomization of a surface loop and subsequent screens (Voss and Skerra 1997). For reactions that need to be quantitated, 96-well microtiter formats can be used. There is currently a vigorous effort underway to increase the number of wells and reduce the required volumes, for pharmaceutical high-throughput screens. Whether this investment in dedicated hardware will be productive, or whether ultimately modular cellular selection systems will supersede screening efforts remains to be seen.

### 3.3.2 Panning

Panning is most frequently employed in the selection of ligand binding molecules, by their enrichment in a pool after binding to a matrix. Thus it can be considered an in vitro selection protocol.

The question of what exactly is being selected for, e.g. by panning a phage display library, deserves some consideration. Depending on the experimental protocol, either equilibrium or dissociation rate constant govern the amount of phage retained on the binding matrix which can subsequently be eluted (Mandecki et al. 1995). Dissociation constants between nanomolar and micromolar can be determined for interactions that have been enriched (Dyson et al. 1995). Obviously, at the upper level, this is far larger than the concentration of phage in the medium. Either multivalence or rebinding affects must be invoked to explain the observed binding of intermediate affinity proteins to the matrix, over the 10–20 washing steps suggested in current protocols (Mccafferty and Johnson 1996), or the selection is in fact for slow dissociation rates. The latter interpretation is corroborated by the finding that the affinity of eluted phages correlates with the time points of collecting them, which can be monitored directly on a surface plasmon resonance chip (Malmborg et al. 1996).

As an alternative to matrix-based panning, fluorescently labeled cells can be sorted directly in a FACS. Again, GFP provides an elegant and modular access to read out a large spectrum of signals that modulate its expression. This was demonstrated with the optimization of GFP itself for FACS sorting. A library of 20 randomized residues flanking the GFP fluorophore yielded variants with 100-fold increased brightness in the cell (Cormack et al. 1996).

Yet another interesting alternative involves exploiting bacterial chemotaxis to select desired functions. This has already been used in a mutational analysis of the *E. coli* chemotaxis receptor Trg (Baumgartner and Hazelbauer 1996). An improved understanding of the molecular mechanisms of bacterial responses to physical and chemical stimuli (Grebe and Stock 1998) can be expected to significantly contribute to the tool kit of available selection systems that can be functionalized.

### 3.3.3 Selecting for Growth

Classically, a system for selection confers a growth advantage on the cell carrying a library molecule with the desired properties. Such systems are conceptually simple and have been widely used; unfortunately they have the highest chance of false positives since a cell generally will have a number of options to cope with selective pressure. This is a rather general phenomenon, for example, an experiment with a T4-lysozyme mutant library under selective pressure to complement a β-galactosidase deficiency was not successful in altering the enzyme's substrate specificity but uncovered a novel *E. coli* locus that weakly complements the defect (Patten et al. 1996).

#### *3.3.3.1 Functional Complementation*

Obviously, functional complementation of a genetic defect will confer a selective advantage in a suitable host strain. This principle has been frequently applied: a recent example is the construction of an active dihydrofolate reductase that is formed from two fragments of the polypeptide when they are brought into proximity by two interacting proteins (Pelletier et al. 1998). A powerful extension of this principle selects proteins in host cells living in extreme environments. This strategy was first used to isolate thermostable variants of kanamycin nucleotidyltransferase, generated in an *E. coli* mutator strain, by transforming a shuttle vector into *Bacillus stearothermophilus* and selecting for growth at elevated temperatures up to 70°C (Liao et al. 1986). Subsequently, further variants were identified and combined and it could be shown that these evolved, thermostable enzymes were at the same time more resistant to protease, urea, detergents and organic solvents (Liao 1993). The same procedure has been applied to chloramphenicol acetyltransferase at 58°C (Turner et al. 1992). An obvious further improvement of this concept would be to use hyperthermophile Archaebacteria- or eubacteria. Unfortunately, molecular biology is a lot more difficult in these cells – the absence of transformable genetic elements requires chromosomal integration for recombinant expression. Nevertheless, the group of T. Oshima has recently developed a shuttle integration vector system for this purpose (Tamakoshi et al. 1997) and successfully used it to stabilize *B. subtilis* isopropylmalate dehydrogenase in *Thermus thermophilus* by gradual adaptation of the integrated gene to growth at up to 70°C in a leuB-deficient strain (Akanuma et al. 1998). Transformation of hyperthermophiles is an active area of research, and progress is under way (Noll and Vargas 1997).

*3.3.3.2 Modular Systems Based on Reporter Genes*

The most universal strategy for selecting novel functions is to couple the function to the expression of some selectable reporter gene, such as antibiotic resistance. A variety of systems has been used in evolutionary engineering projects for this purpose. In eukaryotic cells, the two-hybrid system has made a profound impact with a large number of variations to screen and select for protein–protein interactions (Brachmann and Boeke 1997; Colas and Brent 1998). Prokaryotes have received less attention in this respect – partly because the motivation behind developing the two-hybrid system came from questions in cell biology, partly because eukaryotic transcriptional activation is a particularly intensely studied area. Nonetheless, there are still significant advantages to work with prokaryotes in engineering: transformation numbers are much higher, molecular biology is simpler and growth is faster. A widespread family of bacterial transcriptional regulators is the AraC/XylS protein family of "winged-helix-turn-helix" transcription factors (Gallegos et al. 1997; Martinez and Stock 1997). One member of this family, ToxR of *Vibrio cholerae*, is activated by periplasmic dimerization. Chimeras of the cytoplasmic and transmembrane segment with a periplasmic immunoglobulin domain are functional after dimerization and activate transcription of a reporter gene from the *ctx* promoter (Kolmar et al. 1994, 1995b). This system was successfully used for the screening of stabilized immunoglobulin domains (Kolmar et al. 1995a) and mutational analysis of a dimerizing transmembrane segment (Langosch et al. 1996; Brosig and Langosch 1998) and has recently been further optimized (Jappelli and Brenner 1998).

A different approach to a genetic screen has been developed, based on the phage lambda N protein which induces the modification of *E. coli* RNA polymerase to a termination-resistant form. This anti-termination screen was originally used for the identification and optimization of RNA binding peptides (Harada et al. 1996, 1997), but a generalization appears straightforward.

### 3.3.4 Screening and Selecting Second Site Suppressors

A powerful alternative to searching for mutations that improve a protein may be the search for second site suppressors of a previously introduced deleterious mutation. Whenever a desired property arises from additive effects, a mutation in one site may be compensated for by a sequence change in a different site. The combination of wild-type sequence and second-site suppressor can be expected to improve the protein over and above the wild-type. The advantage of this approach is that baseline activity in the screening experiment can be reduced. This greatly simplifies the detection of successful variants. While the method is general, e.g. it may allow further improvement of enzymes that already function at a level in which a further increase of activity or stability may not be readily detectable, the downside is that not all second site mutations must also improve the wild-type. An early success with this approach was reported for ribonuclease HI, which had been previously destabilized through COOH-terminal deletions (HARUKI et al. 1994). Of 11 second-site suppressor mutations that were identified, eight were also found to improve the wild-type protein. Even thermostable enzymes may be further improved with this strategy, as demonstrated for isopropylmalate dehydrogenase with a chimeric, destabilized enzyme that was subjected to random mutagenesis and selected in an auxotrophic variant of *Thermus thermophilus* at high temperature (KOTSUKA et al. 1996).

In another example, a monomeric variant of chorismate mutase has been engineered by introducing point mutations at the dimer interface. The resulting monomer has almost no detectable enzymatic activity and is significantly destabilized relative to the wild-type. A library of sequences of an interhelical turn was screened for activity, resulting in a variant with almost native catalytic rates (MACBEATH et al. 1998).

### 3.3.5 You Get (Exactly!) What You Ask For

One last caveat may be in order. Evolutionary procedures optimize a fitness function which is not completely under control of the experimenter. For instance, a careful investigation of binding determinants in antibody CDRs of a phage-displayed scFv against fluorescein, using the SIP method, demonstrated that the selection is influenced by a composite fitness function, including affinity, stability and efficient folding (PEDRAZZI et al. 1997). Examples of surprising results exist, like high affinity binders to the column matrix, or enzymes with reduced activities but higher expression levels. The importance of careful experimental design must be emphasized.

## 4 Outlook

The speed and quality of evolutionary solutions to protein engineering problems is truly impressive, ever less knowledge is required about the system that is being optimized. One of the most pointed applications of this principle is the simultaneous engineering of a multigene operon, the arsenate resistance operon of *Staphylococcus aureus* (CRAMERI et al. 1997). While the wild-type plasmid conferred resistance to *E. coli* at a level of 4–10 mM arsenate, after three rounds of DNA shuffling and selection, operons were recovered that conferred resistance up to 400 mM arsenate to the host cells. In addition to ten silent mutations, only three missense mutations in the arsenite membrane pump gene, *arsB*, were sufficient for the increased resistance. Besides improving expression levels and specific activity, apparently an improved functional coupling of the proteins to each other had occurred. This impressive improvement in function, in the absence of a structural model or even a precise understanding of the molecular details of the protein's interactions or the rate-limiting step, is a good indication that evolutionary protein engineering is rapidly moving biotechnology into a new phase.

# References


Airaksinen A, Hovi T (1998) Modified base compositions at degenerate positions of a mutagenic oligonucleotide enhance randomness in site-saturation mutagenesis. Nucleic Acids Res 26:576–581

Akanuma S, Yamagishi A, Tanaka N, Oshima T (1998) Serial increase in the thermal stability of 3-isopropylmalate dehydrogenase from *Bacillus subtilis* by experimental evolution. Protein Sci 7:698–705

Atwell S, Ridgway JB, Wells JA, Carter P (1997) Stable heterodimers from remodeling the domain interface of a homodimer using a phage display library. J Mol Biol 270:26–35

Aurora R, Rose GD (1998) Helix capping. Protein Sci 7:21–38

Baca M, Scanlan TS, Stephenson RC, Wells JA (1997) Phage display of a catalytic antibody to optimize affinity for transition-state analog binding. Proc Natl Acad Sci USA 94:10063–10068

Bachmann MF, Kundig TM, Kalberer CP, Hengartner H, Zinkernagel RM (1994) How many specific B cells are needed to protect against a virus? J Immunol 152:4235–4241

Balint RF, Larrick JW (1993) Antibody engineering by parsimonious mutagenesis. Gene 137:109–118

Barbas C, Heine A, Zhong G, Hoffmann T, Gramatikova S, Bjornestedt R, List B, Anderson J, Stura EA, Wilson IA, Lerner RA (1997) Immune versus natural selection: antibody aldolases with enzymic rates but broader scope. Science 278:2085–2092

Baumgartner JW, Hazelbauer GL (1996) Mutational analysis of a transmembrane segment in a bacterial chemoreceptor. J Bacteriol 178:4651–4660

Beste G, Schmidt FS, Stibora T, Skerra A (1999) Small antibody-like proteins with prescribed ligand specificities derived from the lipocalin fold. Proc Natl Acad Sci USA 96:1898–1903

Borchardt A, Liberles SD, Biggar SR, Crabtree GR, Schreiber SL (1997) Small molecule-dependent genetic selection in stochastic nanodroplets as a means of detecting protein-ligand interactions on a large scale. Chem Biol 4:961–968

Bothmann H, Plückthun A (1998) Selection for a periplasmic factor improving phage display and functional periplasmic expression. Nature Biotech 16:376–380

Brachmann RK, Boeke JD (1997) Tag games in yeast: the two-hybrid system and beyond. Curr Op Biotech 8:561–568

Brosig B, Langosch D (1998) The dimerization motif of the glycophorin A transmembrane segment in membranes: importance of glycine residues. Protein Sci 7:1052–1056

Buchholz F, Angrand PO, Stewart AF (1998) Improved properties of FLP recombinase evolved by cycling mutagenesis. Nature Biotech 16:657–662

Burton DR (1995) Phage display. Immunotech 1:87–94

Cadwell RC, Joyce GF (1994) Mutagenic PCR. PCR Methods Appl 3:S136–140

Carter P, Wells JA (1988) Dissecting the catalytic triad of a serine protease. Nature 332:564–568

Cech TR (1990) Self-splicing of group I introns. Annu Rev Biochem 59:543–568

Chen R, Henning U (1996) A periplasmic protein (Skp) of Escherichia coli selectively binds a class of outer membrane proteins. Mol Microbiol 19:1287–1294

Chien CT, Bartel PL, Sternglanz R, Fields S (1991) The two-hybrid system: a method to identify and clone genes for proteins that interact with a protein of interest. Proc Natl Acad Sci USA 88:9578–9582

Colas P, Brent R (1998) The impact of two-hybrid and related methods on biotechnology. Trends Biotech 16:355–363

Cormack BP, Valdivia RH, Falkow S (1996) FACS-optimized mutants of the green fluorescent protein (GFP). Gene 33–38

Cowan DA (1997) Thermophilic proteins: stability and function in aqueous and organic solvents. Comp Biochem Physiol A Physiol 118:429–438

Crameri A, Dawes G, Rodriguez EJ, Silver S, Stemmer WP (1997) Molecular evolution of an arsenate detoxification pathway by DNA shuffling. Nature Biotech 15:436–438

Crameri A, Raillard SA, Bermudez E, Stemmer WP (1998) DNA shuffling of a family of genes from diverse species accelerates directed evolution. Nature 391:288–291

Crameri A, Whitehorn EA, Tate E, Stemmer WP (1996) Improved green fluorescent protein by molecular evolution using DNA shuffling. Nature Biotech 14:315–319

Cull MG, Miller JF, Schatz PJ (1992) Screening for receptor ligands using large libraries of peptides linked to the C terminus of the *lac* repressor. Proc Natl Acad Sci USA 89:1865–1869

Cunningham BC, Wells JA (1987) Improvement in the alkaline stability of subtilisin using an efficient random mutagenesis and screening procedure. Protein Eng 1:319–325

Czarnik AW (1997) Encoding strategies in combinatorial chemistry. Proc Natl Acad Sci USA 94: 12738–12739

Delagrave S, Goldman ER, Youvan DC (1993) Recursive ensemble mutagenesis. Protein Eng 6:327–331

Dyson MR, Germaschewski V, Murray K (1995) Direct measurement via phage titre of the dissociation constants in solution of fusion phage-substrate complexes. Nucleic Acids Res 23:1531–1535

Eigen M, Rigler R (1994) Sorting single molecules: application to diagnostics and evolutionary biotechnology. Proc Natl Acad Sci USA 91:5740–5747

Fisch I, Kontermann RE, Finnern R, Hartley O, Soler GA, Griffiths AD, Winter G (1996) A strategy of exon shuffling for making large peptide repertoires displayed on filamentous bacteriophage. Proc Natl Acad Sci USA 93:7761–7766

Fuchs P, Weichel W, Dübel S, Breitling F, Little M (1996) Separation of *E. coli* expressing functional cell-wall bound antibody fragments by FACS. Immunotech 2:97–102

Gallegos MT, Schleif R, Bairoch A, Hofmann K, Ramos JL (1997) Arac/XylS family of transcriptional regulators. Microbiol Mol Biol Rev 61:393–410

Gates CM, Stemmer WP, Kaptein R, Schatz PJ (1996) Affinity selective isolation of ligands from peptide libraries through display on a *lac* repressor "headpiece dimer". J Mol Biol 255:373–386

Gaytan P, Yanez J, Sanchez F, Mackie H, Soberon X (1998) Combination of DMT-mononucleotide and Fmoc-trinucleotide phosphoramidites in oligonucleotide synthesis affords an automatable codon-level mutagenesis method. Chem Biol 5:519–527

Georgiou G, Stathopoulos C, Daugherty PS, Nayak AR, Iverson BL, Curtiss Rr (1997) Display of heterologous proteins on the surface of microorganisms: from the screening of combinatorial libraries to live recombinant vaccines. Nature Biotech 15:29–34

Grebe TW, Stock J (1998) Bacterial chemotaxis: the five sensors of a bacterium. Curr Biol 8:R154–R157

Griffiths AD, Duncan AR (1998) Strategies for selection of antibodies by phage display. Curr Op Biotech 9:102–108

Groskreutz D, Schenborn ET (1997) Reporter systems. Meth Mol Biol 63:11–30

Hanes J, Plückthun A (1997) In vitro selection and evolution of functional proteins by using ribosome display. Proc Natl Acad Sci USA 94:4937–4942

Harada K, Martin SS, Frankel AD (1996) Selection of RNA-binding peptides in vivo. Nature 380: 175–179

Harada K, Martin SS, Tan R, Frankel AD (1997) Molding a peptide into an RNA site by in vivo peptide evolution. Proc Natl Acad Sci USA 94:11887–11892

Haruki M, Noguchi E, Akasako A, Oobatake M, Itaya M, Kanaya S (1994) A novel strategy for stabilization of *Escherichia coli* ribonuclease HI involving a screen for an intragenic suppressor of carboxyl-terminal deletions. J Biol Chem 269:26904–26911

He M, Taussig MJ (1997) Antibody-ribosome-mRNA (ARM) complexes as efficient selection particles for in vitro display and evolution of antibody combining sites. Nucleic Acids Res 25:5132–5134

Hedstrom L (1996) Trypsin: a case study in the structural determinants of enzyme specificity. Biol Chem 377:465–470

Hennecke F, Krebber C, Plückthun A (1998) Non-repetitive single-chain Fv linkers selected by selectively infective phage (SIP) technology. Protein Eng 11:405–410

Hoess R, Abremski K, Sternberg N (1984) The nature of the interaction of the P1 recombinase Cre with the recombining site *loxP*. Cold Spring Harb Symp Quant Biol 49:761–768

Hoogenboom HR, de BA, Hufton SE, Hoet RM, Arends JW, Roovers RC (1998) Antibody phage display technology and its applications. Immunotech 4:1–20

Janda KD, Lo LC, Lo C, Sim MM, Wang R, Wong CH, Lerner RA (1997) Chemical selection for catalysis in combinatorial antibody libraries. Science 275:945–948

Jappelli R, Brenner S (1998) Changes in the periplasmic linker and in the expression level affect the activity of ToxR and lambda-ToxR fusion proteins in *Escherichia coli*. FEBS Lett 423:371–375

Jencks WP (1969) Catalysis in Chemistry and Enzymology, Dover, Mineola, NY

Johnson CM, Oliveberg M, Clarke J, Fersht AR (1997) Thermodynamics of denaturation of mutants of barnase with disulfide crosslinks. J Mol Biol 268:198–208

Joyet P, Declerck N, Gaillardin C (1992) Hyperthermostable variants of a highly thermostable alpha-amylase. Biotechnology (N Y) 10:1579–1583

Kauffman SA (1993) The Origins of Order. Self-organization and Selection in Evolution, Oxford University Press, New York, Oxford

Kauffman SA, Macready WG (1995) Search strategies for applied molecular evolution. J Theor Biol 173:427–440

Kayushin AL, Korosteleva MD, Miroshnikov AI, Kosch W, Zubov D, Piel N (1996) A convenient approach to the synthesis of trinucleotide phosphoramidites-synthons for the generation of oligonucleotide/peptide libraries. Nucleic Acids Res 24:3748–3755

Kiefhaber T, Grunert HP, Hahn U, Schmid FX (1990) Replacement of a *cis* proline simplifies the mechanism of ribonuclease T1 folding. Biochemistry 29:6475–6480

Kiefhaber T, Rudolph R, Kohler HH, Buchner J (1991) Protein aggregation in vitro and in vivo: a quantitative model of the kinetic competition between folding and aggregation. Biotechnology (N Y) 9:825–829

Knappik A, Plückthun A (1995) Engineered turns of a recombinant antibody improve its in vivo folding. Protein Eng 8:81–89

Kolmar H, Frisch C, Gotze K, Fritz HJ (1995a) Immunoglobulin mutant library genetically screened for folding stability exploiting bacterial signal transduction. J Mol Biol 251:471–476

Kolmar H, Frisch C, Kleemann G, Gotze K, Stevens FJ, Fritz HJ (1994) Dimerization of Bence Jones proteins: linking the rate of transcription from an *Escherichia coli* promoter to the association constant of REIV. Biol Chem Hoppe Seyler 375:61–70

Kolmar H, Hennecke F, Gotze K, Janzer B, Vogt B, Mayer F, Fritz HJ (1995b) Membrane insertion of the bacterial signal transduction protein ToxR and requirements of transcription activation studied by modular replacement of different protein substructures. EMBO J 14:3895–3904

Kotsuka T, Akanuma S, Tomuro M, Yamagishi A, Oshima T (1996) Further stabilization of 3-isopropylmalate dehydrogenase of an extreme thermophile, *Thermus thermophilus*, by a suppressor mutation method. J Bacteriol 178:723–727

Krebber A, Burmester J, Plückthun A (1996) Inclusion of an upstream transcriptional terminator in phage display vectors abolishes background expression of toxic fusions with coat protein g3p. Gene 178:71–74

Kumamaru T, Suenaga H, Mitsuoka M, Watanabe T, Furukawa K (1998) Enhanced degradation of polychlorinated biphenyls by directed evolution of biphenyl dioxygenase. Nature Biotech 16:663–666

Lam KS, Salmon SE, Hersh EM, Hruby VJ, Kazmierski WM, Knapp RJ (1991) A new type of synthetic peptide library for identifying ligand-binding activity. Nature 354:82–84

Langosch D, Brosig B, Kolmar H, Fritz HJ (1996) Dimerisation of the glycophorin A transmembrane segment in membranes probed with the ToxR transcription activator. J Mol Biol 263:525–530

Lerner RA, Benkovic SJ, Schultz PG (1991) At the crossroads of chemistry and immunology: catalytic antibodies. Science 252:659–657

Leung D, Chen E, Goeddel D (1989) A method for Random Mutagenesis of a defined DNA segment using a Modified Polymerase Chain Reaction. Technique 1:11–15

Liao H, McKenzie T, Hageman R (1986) Isolation of a thermostable enzyme variant by cloning and selection in a thermophile. Proc Natl Acad Sci USA 83:576–580

Liao HH (1993) Thermostable mutants of kanamycin nucleotidyltransferase are also more stable to proteinase K, urea, detergents, and water-miscible organic solvents. Enzyme Microb Technol 15:286–292

Light J, Lerner RA (1995) Random mutagenesis of staphylococcal nuclease and phage display selection. Bioorg Med Chem 3:955–967

Lyttle MH, Napolitano EW, Calio BL, Kauvar LM (1995) Mutagenesis using trinucleotide beta-cyanoethyl phosphoramidites. Biotechniques 19:274–281

MacBeath G, Kast P, Hilvert D (1998) Redesigning enzyme topology by directed evolution. Science 279:1958–1961

Makowski L (1994) Phage display: structure, assembly and engineering of filamentous bacteriophage M13. Curr Opin Struct Biol 4:225–230

Malakauskas SM, Mayo SL (1998) Design, structure and stability of a hyperthermophilic protein variant. Nature Struct Biol 5:470–475

Malmborg AC, Duenas M, Ohlin M, Soderlind E, Borrebaeck CA (1996) Selection of binders from phage displayed antibody libraries using the BIAcore biosensor. J Immunol Methods 198:51–57

Malmborg AC, Soderlind E, Frost L, Borrebaeck CA (1997) Selective phage infection mediated by epitope expression on F pilus. J Mol Biol 273:544–551

Mandecki W, Chen YC, Grihalde N (1995) A mathematical model for biopanning (affinity selection) using peptide libraries on filamentous phage. J Theor Biol 176:523–530

Manen D, Pougeon M, Damay P, Geiselmann J (1997) A sensitive reporter gene system using bacterial luciferase based on a series of plasmid cloning vectors compatible with derivatives of pBR322. Gene 186:197–200

Markland W, Ley AC, Lee SW, Ladner RC (1996) Iterative optimization of high-affinity proteases inhibitors using phage display. 1. Plasmin. Biochemistry 35:8045–8057
Martineau P, Jones P, Winter G (1998) Expression of an antibody fragment at high levels in the bacterial cytoplasm. J Mol Biol 280:117–127
Martinez HE, Stock AM (1997) Structural relationships in the OmpR family of winged-helix transcription factors. J Mol Biol 269:301–312
Mattheakis LC, Dias JM, Dower WJ (1996) Cell-free synthesis of peptide libraries displayed on polysomes. Meth Enz 267:195–207
McCafferty J, Jackson RH, Chiswell DJ (1991) Phage-enzymes: expression and affinity chromatography of functional alkaline phosphatase on the surface of bacteriophage. Protein Eng 4:955–961
McCafferty J, Johnson KS (1996) Construction and screening of antibody display libraries. Phage display of peptides and proteins. A laboratory manual (Kay, B. K., Winter, J. & McCafferty, J., Eds.), Academic Press, San Diego.
Miesenböck G, DeAngelis AD, Rothman JE (1998) Visualizing secretion and synaptic transmission with pH-sensitive green fluorescent proteins. Nature 394:192–195
Miyawaki A, Llopis J, Heim R, McCaffery JM, Adams JA, Ikura M, Tsien RY (1997) Fluorescent indicators for $Ca2+$ based on green fluorescent proteins and calmodulin. Nature 388:882–887
Moore JC, Jin HM, Kuchner O, Arnold FH (1997) Strategies for the in vitro evolution of protein function: enzyme evolution by random recombination of improved sequences. J Mol Biol 272:336–347
Needels MC, Jones DG, Tate EH, Heinkel GL, Kochersperger LM, Dower WJ, Barrett RW, Gallop MA (1993) Generation and screening of an oligonucleotide-encoded synthetic peptide library. Proc Natl Acad Sci USA 90:10700–10704
Nemoto N, Miyamoto SE, Husimi Y, Yanagawa H (1997) In vitro virus: bonding of mRNA bearing puromycin at the 3′-terminal end to the C-terminal end of its encoded protein on the ribosome in vitro. FEBS Lett 414:405–408
Nicholson H, Tronrud DE, Becktel WJ, Matthews BW (1992) Analysis of the Effectiveness of Proline Substitutions and Glycine Replacements in Increasing the Stability of Phage T4 Lysozyme. Biopolymers 32:1431–1441
Noll KM, Vargas M (1997) Recent advances in genetic analyses of hyperthermophilic archaea and bacteria. Arch Microbiol 168:73–80
Oberholzer T, Albrizio M, Luisi PL (1995) Polymerase chain reaction in liposomes. Chem Biol 2:677–682
Ohage EC, Graml W, Walter MM, Steinbacher S, Steipe B (1997) b-Turn propensities as paradigms for the analysis of structural motifs to engineer protein stability. Protein Sci 6:233–241
Oliphant AR, Nussbaum AL, Struhl K (1986) Cloning of random-sequence oligodeoxynucleotides. Gene 44:177–183
Ono A, Matsuda A, Zhao J, Santi DV (1995) The synthesis of blocked triplet-phosphoramidites and their use in mutagenesis. Nucleic Acids Res 23:4677–4682
Pai LH, Wittes R, Setser A, Willingham MC, Pastan I (1996) Treatment of advanced solid tumors with immunotoxin LMB-1: an antibody linked to Pseudomonas exotoxin. Nature Med 2:350–353
Patten PA, Sonoda T, Davis MM (1996) Directed evolution studies with combinatorial libraries of T4 lysozyme mutants. Mol Divers 1:97–108
Pedrazzi G, Schwesinger F, Honegger A, Krebber C, Plückthun A (1997) Affinity and folding properties both influence the selection of antibodies with the selectively infective phage (SIP) methodology. FEBS Lett 415:289–293
Pelletier JN, Campbell VF, Michnick SW (1998) Oligomerization domain-directed reassembly of active dihydrofolate reductase from rationally designed fragments. Proc Natl Acad Sci USA 95:12141–12146
Pini A, Viti F, Santucci A, Carnemolla B, Zardi L, Neri P, Neri D (1998) Design and use of a phage display library. Human antibodies with subnanomolar affinity against a marker of angiogenesis eluted from a two-dimensional gel. J Biol Chem 273:21769–21776
Rebar EJ, Pabo CO (1994) Zinc finger phage: affinity selection of fingers with new DNA-binding specificities. Science 263:671–673
Roberts RW, Szostak JW (1997) RNA-peptide fusions for the in vitro selection of peptides and proteins. Proc Natl Acad Sci USA 94:12297–12302
Rubingh DN (1997) Protein engineering from a bioindustrial point of view. Curr Op Biotech 8:417–422
Rudolph R, Lilie H (1996) In vitro folding of inclusion body proteins. FASEB J 10:49–56
Schatz PJ, Cull MG, Martin EL, Gates CM (1996) Screening of peptide libraries linked to lac repressor. Meth Enz 267:171–191
Schmid FX, Frech C, Scholz C, Walter S (1996) Catalyzed and assisted protein folding of ribonuclease T1. Biol Chem 377:417–424

Shafikhani S, Siegel RA, Ferrari E, Schellenberger V (1997) Generation of large libraries of random mutants in *Bacillus subtilis* by PCR-based plasmid multimerization. Biotechniques 23:304–310
Shao Z, Zhao H, Giver L, Arnold FH (1998) Random-priming in vitro recombination: an effective tool for directed evolution. Nucleic Acids Res 26:681–683
Sidhu SS, Borgford TJ (1996) Selection of *Streptomyces griseus* protease B mutants with desired alterations in primary specificity using a library screening strategy. J Mol Biol 257:233–245
Singer B, Kusmierek JT (1982) Chemical mutagenesis. Annu Rev Biochem 51:655–693
Skerra A (1994) Use of the tetracycline promoter for the tightly regulated production of a murine antibody fragment in *Escherichia coli*. Gene 151:131–135
Soderlind E, Lagerkvist AC, Duenas M, Malmborg AC, Ayala M, Danielsson L, Borrebaeck CA (1993) Chaperonin assisted phage display of antibody fragments on filamentous bacteriophages. Biotechnology (N Y) 11:503–507
Soumillion P, Jespers L, Bouchet M, Marchand BJ, Winter G, Fastrez J (1994) Selection of β-lactamase on filamentous bacteriophage by catalytic activity. J Mol Biol 237:415–422
Spada S, Krebber C, Plückthun A (1997) Selectively infective phages (SIP). Biol Chem 378:445–456
Steipe B, Schiller B, Plückthun A, Steinbacher S (1994) Sequence Statistics Reliably Predict Stabilizing Mutations in a Protein Domain. J Mol Biol 240:188–192
Stemmer WP (1994a) DNA shuffling by random fragmentation and reassembly: in vitro recombination for molecular evolution. Proc Natl Acad Sci USA 91:10747–10751
Stemmer WP (1994b) Rapid evolution of a protein in vitro by DNA shuffling. Nature 370:389–391
Sternberg N, Hoess RH (1995) Display of peptides and proteins on the surface of bacteriophage lambda. Proc Natl Acad Sci USA 92:1609–1613
Strausberg SL, Alexander PA, Gallagher DT, Gilliland GL, Barnett BL, Bryan PN (1995) Directed evolution of a subtilisin with calcium-independent stability. Biotechnology (N Y) 13:669–673
Tamakoshi M, Uchida M, Tanabe K, Fukuyama S, Yamagishi A, Oshima T (1997) A new *Thermus-Escherichia coli* shuttle integration vector system. J Bacteriol 179:4811–4814
Tawfik DS, Griffiths AD (1998) Man-made cell-like compartments for molecular evolution. Nature Biotech 16:652–656
Tsien RY (1998) The green fluorescent protein. Annu Rev Biochem 67:509–544
Turner SL, Ford GC, Mountain A, Moir A (1992) Selection of a thermostable variant of chloramphenicol acetyltransferase (Cat-86). Protein Eng 5:535–541
Van den Burg B, Vriend G, Veltman OR, Venema G, Eijsink VG (1998) Engineering an enzyme to resist boiling. Proc Natl Acad Sci USA 95:2056–2060
Vaughan TJ, Williams AJ, Pritchard K, Osbourn JK, Pope AR, Earnshaw JC, McCafferty J, Hodits RA, Wilton J, Johnson KS (1996) Human antibodies with sub-nanomolar affinities isolated from a large non-immunized phage display library. Nature Biotech 14:309–314
Virnekäs B, Ge L, Plückthun A, Schneider KC, Wellnhofer G, Moroney SE (1994) Trinucleotide phosphoramidites: ideal reagents for the synthesis of mixed oligonucleotides for random mutagenesis. Nucleic Acids Res 22:5600–5607
Vispo NS, Callejo M, Ojalvo AG, Santos A, Chinea G, Gavilondo JV, Arana MJ (1997) Displaying human interleukin-2 on the surface of bacteriophage. Immunotech 3:185–193
Voss S, Skerra A (1997) Mutagenesis of a flexible loop in streptavidin leads to higher affinity for the Strep-tag II peptide and improved performance in recombinant protein purification. Protein Eng 10:975–982
Walter S, Hubner B, Hahn U, Schmid FX (1995) Destabilization of a protein helix by electrostatic interactions. J Mol Biol 252:133–143
Wang CI, Yang Q, Craik CS (1996) Phage display of proteases and macromolecular inhibitors. Meth Enz 267:52–68
Wentworth P, Janda KD (1998) Catalytic antibodies. Curr Op Chem Biol 2:138–144
Wirsching P, Ashley JA, Lo CH, Janda KD, Lerner RA (1995) Reactive immunization. Science 270:1775–1782
You L, Arnold FH (1996) Directed evolution of subtilisin E in *Bacillus subtilis* to enhance total activity in aqueous dimethylformamide. Protein Eng 9:77–83
Zhang X-J, Baase WA, Matthews BW (1992) Multiple alanine replacements within a-helix 126–134 of T4 lysozyme have independent, additive effects on both structure and stability. Protein Sci 1:761–776
Zhang X-j, Baase WA, Shoichet BK, Wilson KP, Matthews BW (1995) Enhancement of protein stability by the combination of point mutations in T4 lysozyme is additive. Protein Eng 8:1017–1022
Zhao H, Arnold FH (1997) Optimization of DNA shuffling for high fidelity recombination. Nucleic Acids Res 25:1307–1308

Zhao H, Giver L, Shao Z, Affholter JA, Arnold FH (1998) Molecular evolution by staggered extension process (StEP) in vitro recombination. Nature Biotech 16:258–261

Zlokarnik G, Negulescu PA, Knapp TE, Mere L, Burres N, Feng L, Whitney M, Roemer K, Tsien RY (1998) Quantitation of transcription and clonal selection of single living cells with β-lactamase as reporter. Science 279:84–88

# Phage Display of Combinatorial Peptide and Protein Libraries and Their Applications in Biology and Chemistry

K. Johnsson[1] and L. Ge[2]



## 1 Introduction

The enormous potential of combinatorial approaches for studying problems in biology and chemistry has been clearly demonstrated over the last several years. In particular, phage display has become one of the major techniques for the use of combinatorial peptide and protein libraries. The reasons for the success of phage display are, at least, threefold: (1) phage display of combinatorial libraries creates a direct link between phenotype (the selectable properties of interest from the displayed library) and genotype (their sequences); (2) it offers the possibility to select and amplify single clones out of large libraries; (3) it allows in vitro as well as in vivo selections in order to evolve peptides and proteins with novel activities. Reported applications include selections of peptides as lead compounds in phar-

[1] Lehrstuhl für Organische Chemie I, Bioorganische Chemie, Ruhr-Universität Bochum, D-44780 Bochum, Germany
[2] XERION PHARMACEUTICALS GmbH, Fraunhoferstr. 9, D-82152 Martinsried, Germany

maceutical research, redesign of protein structures and protein–protein interactions, screening of cDNA libraries and enzyme design. This diversity in applications has been made possible by the development of a variety of display formats and selection schemes and an increasing understanding of the biology of filamentous bacteriophages.

This review aims to provide the reader with a general overview of the tremendous possibilities of the technology as well as its limitations. Selected examples will be used to highlight the current state of the art, although these examples represent only a fraction of all the interesting experiments and techniques that have been developed over the last few years.

## 2 Filamentous Bacteriophage

The most commonly used bacteriophages for phage display of proteins or peptides are derivatives of F-specific filamentous bacteriophages, fd, f1 or M13. They infect male ($F^+$) strains of *Escherichia coli* and belong to the best studied group of bacteriophages in terms of their physiology and genetics. For an extensive overview of the biology of filamentous bacteriophages, readers are referred to the excellent review by Model and Russel (1988). Only a brief review of the properties of filamentous bacteriophages and their relevance to phage display will be presented here. Clearly, a detailed understanding of host (bacteria) and phage biology is important for an appreciation of the possibilities and limitations of phage display.

The most distinguishing characteristics of filamentous phages include the variable length of the phage capsid, which is determined by the size of single-stranded DNA encapsidated inside, and the continuous production/extrusion from the bacteria without lysis of their hosts. Five phage proteins (gene III, VI, VII, VIII and IX proteins) are involved in virion capsid formation. Gene VIII protein (gVIIIp) is the major coat protein, present in approximately 2700 copies and comprising the sides of the thin, rod-like virus. However, the precise number of gVIIIp is determined by the size of encapsidated single-stranded DNA. There are approximately five copies of each gene III and gene VI proteins (gIIIp and gVIp, see Goldsmith and Konigsberg 1977; Lin et al. 1980) located at one end of the virion. Mature gIIIp consists of three domains, a COOH-terminal domain, responsible for anchoring gIIIp in the phage particle, and two $NH_2$-terminal domains, responsible for infecting *E. coli*. At the opposite end of the phage, five copies of gene VII (gVIIp) and gene IX (gIXp) proteins are required for stabilization and termination of the virion (Grant et al. 1981).

Of the five capsid proteins, only gIIIp and gVIIIp are synthesized as precursors containing signal sequences. After synthesis, gVIIIp is rapidly incorporated in the cytoplasmic membrane, with the $NH_2$-terminus exposed in the periplasm and the COOH-terminus on the cytoplasmic side (Wickner 1975; Ohkawa and Webster 1981). Much less is known about the gIIIp membrane insertion. However, it has been shown that the mature protein is held in the cytoplasmic membrane by a membrane-spanning region near its COOH-terminal (the same region responsible for the anchoring of gIIIp through gVIp in the virion capsid) and the rest of the protein is exposed in the periplasm (DAVIS et al. 1985). Recently, it has been demonstrated that other viral structural proteins are also associated with the inner membrane of *E. coli* (ENDEMANN and MODEL 1995).

Assembly of phages occurs when single-stranded (ss) phage genomic DNA or plasmid DNA containing the phage origin of replication and packaging signal (phagemid), together with the capsid proteins, extrudes through the cytoplasmic membrane directly to the cell exterior. The end of the virion terminated by gVIIp and gIXp, together with the packaging signal part of the ssDNA, emerges first (LOPEZ and WEBSTER 1983). This is followed by the continuous replacement of the gene V protein, which binds to the nascent ssDNA in the cytoplasm, by the membrane associated gVIIIp. The assembly process is terminated by the emergence of the gIIIp and gVIp end of the phage (LOPEZ and WEBSTER 1983). Since phage display systems are based on fusion to gIIIp, gVIIIp or gVIp, it is advisable to consider if the protein to be displayed is compatible with the production and secretion of the virion capsid proteins.

Infection of host bacteria by phage particles is initiated by docking of the second $NH_2$-terminal domain of gIIIp to the tip of the F pilus of male bacteria (GRAY et al. 1981), followed by the retraction of the pilus to bring the phage particles close to the host cell surface (JACOBSON 1972). Penetration of the phage particles occurs upon binding of the first $NH_2$-terminal domain of gIIIp to the COOH-terminal domain of its host receptor, the TolA protein (RIECHMANN and HOLLIGER 1997), and proceeds through decoating of capsid proteins, which become associated with the inner membrane (SMILOWITZ 1974). Recently, the structure of the two $NH_2$-terminal domains of gIIIp has been solved (HOLLIGER and RIECHMANN 1997; LUBKOWSKI et al. 1998). As these domains are responsible for the docking of the phage particle to the F pilus and the binding to TolA, the structure provides a basis for an understanding of the protein–protein interactions that take place during the infection process.

# 3 Display of Proteins and Peptides on Phage

## 3.1 Filamentous Bacteriophage Display Systems

Foreign proteins or peptides are displayed on filamentous phages through their fusion to minor coat proteins gIIIp (SMITH 1985), gVIp (JESPERS et al. 1995) and major coat protein gVIIIp (MARKLAND et al. 1991; KANG et al. 1991).

The fusion of foreign peptides or proteins to the major coat protein gVIIIp offers the opportunity for multivalent display (Fig. 1). For instance, peptide libraries can be inserted into the $NH_2$-terminal region of gVIIIp and displayed in approximately

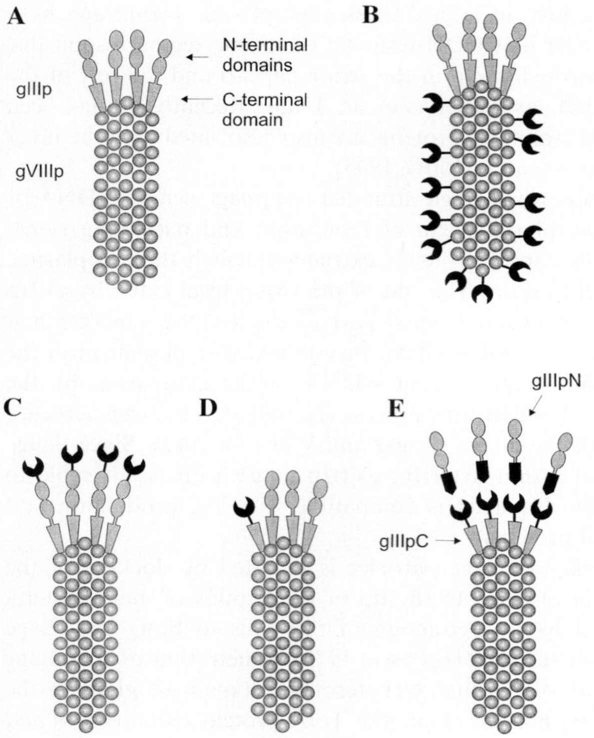


**Fig. 1A–E.** Most commonly used phage display formats. For clarity, only gIIIp and gVIIIp are shown; displayed foreign proteins are shown in *black*. **A** Wild-type filamentous bacteriophage. **B** Multivalent gVIIIp phage display. **C** Multivalent gIIIp display. **D** Monovalent gIIIp phage display; the foreign protein is fused to the COOH-terminal domain of gIIIp. **E** "Selectively infective phage," the interaction between the displayed protein and its target (both shown in *black*) mediates infectivity

2700 copies per phage particle. However, it was shown (Iannolo et al. 1995) that the fraction of viable phages from a peptide library dropped as the length of the peptides increased. So whereas most peptides of six residues were tolerated by the displaying phages, only 1% of the phages containing inserted peptides of 16 residues could be recovered. Using phagemids, in which the necessary phage proteins are supplied in *trans* by helper phages, proteins as large as antibody Fab fragments can be displayed and selected as gVIIIp fusion (Kang et al. 1991). Depending on the peptide or protein displayed, the copy number of gVIIIp fusion proteins in these hybrid virions ranges from 1000 to less than one per phage. The exact reasons for these differences are not well understood. However, a correlation between processing of pro-coat gVIIIp fusion protein (insertion into the membrane and processing by leader peptidase) and the number of mature gVIIIp fusion proteins in the hybrid virion has recently been demonstrated (Malik et al. 1996).

There are few reported experiments available on gVIp based phage display, although in one report it has already been shown that fusion of foreign proteins to the COOH-terminus of gVIp can have negative effects on phage viability (Jespers et al. 1995).

So far the most commonly used phage display format relies on the fusion of foreign peptides or proteins to gIIIp (Fig. 1). Considering the pivotal role played by gIIIp in the phage assembly and infection process, it is quite remarkable that phages displaying a protein or a peptide fused to the $NH_2$-terminus of complete gIIIp retain their ability to assemble and to infect their *E. coli* hosts. Using phagemids, proteins or peptides can also be fused to truncated gIIIp, where the first two domains of gIIIp (necessary for infection, but not for assembly) are deleted (Garrard et al. 1991; Barbas et al. 1991). In this case, wild-type gIIIp has to be provided in *trans* by helper phages to render the displaying phage infectious. Usually, only a small portion of the phage population (less than one in ten) actually displays the fusion protein or peptide (Bass et al. 1991). As correct folding of the gIIIp fusion proteins or peptides in the periplasm is a prerequisite for the functional display of foreign proteins, folding may also play an important role in the copy number of gIIIp fusion proteins. However, this has not been systematically investigated. Very little is known about the factors which influence the folding of proteins transported to the periplasm. Over-expression of some known folding catalysts in the periplasm generated little improvement in periplasmic protein folding (Knappik et al. 1993). An ingenious approach to this problem has been taken by Bothmann and Plückthun, who used an *E. coli* genomic DNA library to select for factor(s) that improved the folding of proteins displayed on phage (Bothmann and Plückthun 1998). A factor selected by this process turned out to be *Skp*, which has previously been suggested to play a role in the transport or folding of outer-membrane proteins (Chen and Henning 1996). Clearly, protein export and folding plays a critical role in all phage display formats.

Although an incorrectly folded protein or peptide is unlikely to be selected, it cannot be assumed that all correctly folded clones can be selected. Notwithstanding the affinity, the selected fusion phages can still be lost during the elution or phage recovery process. It has been observed that fusion phages displaying an antibody scFv fragment lost their viability dramatically during the normal elution step of a selection (10min in 10mM triethylamine), whereas the viability of wild-type helper phages was not affected at all (L. Ge, unpublished results). Other alternative phage recovery processes, for instance acid elution or DTT treatment (which has no effect on wild-type phages, see Rakonjac and Model 1994), may have adverse effects on some of the displayed proteins or peptides, a fact that cannot be predicted in advance.

Regardless of which display format is employed, phage selections often suffer from the amplification of nonspecific binding molecules. A system that directly couples the binding of a displayed peptide or protein (i.e. ligand) to its target with the amplification of the displaying phage could alleviate this problem. Indeed, such a system has independently been developed by three groups (Duenas and Borrebaeck 1994; Gramatikoff et al. 1994; Krebber et al. 1995). Here, the $NH_2$- and COOH-terminal domains of gIIIp (gIIIpN and gIIIpC) are expressed separately, fused to either the ligand (gIIIpC) or its target (gIIIpN) (Fig. 1). Phages displaying only the gIIIpC fusion protein can form stable virions but are noninfectious due to the absence of the gIIIp $NH_2$-terminal domains. However, interaction between target and ligand links the gIIIpN and gIIIpC domains and infectivity is restored, resulting in "selectively infective phage". Consequently, phage propagation strictly depends on target-ligand interaction. This selection process can be accomplished in vivo or in vitro. In vivo, both the gIIIpN and gIIIpC fusion proteins are encoded on the same vector and functional gIIIp forms in the periplasm. gIIIpN can also be expressed separately and fused either genetically or chemically to the corresponding target (Krebber et al. 1997). The gIIIpN fusion protein is subsequently mixed in vitro with the corresponding phage displaying the gIIIpC fusion protein. Excellent specificities and enrichment factors have been reported for this "selectively infective phage" technology.

An alternative system, which also links the binding event between a ligand and its cognate target to the amplification of the phage, has been developed by Borrebaeck and co-workers (Malmborg et al. 1997). In this system an antigenic peptide is expressed as a fusion to F pilin, the building block of the F pilus. Bacteria displaying these modified F pili are resistant to infection by wild-type M13 phage. However, phages displaying a specific antibody to the peptide can infect bacteria displaying the modified F pili, demonstrating that the specific antibody-antigen interaction mediates infection. Enrichment factors in model selections have been around 2500.

## 3.2 Phage Display Systems Other Than Filamentous Phage

The display of peptides and proteins is not limited to filamentous phages. Foreign polypeptides have been displayed on bacteriophage $\lambda$ (Dunn 1996 and references therein), T4 (Jiang et al. 1997; Ren et al. 1996; Efimov et al. 1995), P4 (Lindqvist and Naderi 1995), bacteria (Stahl and Uhlen 1997 and references therein) as well as eukaryotic cells (Boder and Wittrup 1997) and viruses (Boublik et al. 1995). Depending on the protein to be displayed, alternative systems can offer certain advantages over the filamentous phage display system. For example, eukaryotic cells should in general be better suited for the expression of eukaryotic proteins, as they possess the machinery required for proper folding and post-translational modifications. In particular, as a complementary system to filamentous phage display, the display of peptides and proteins on bacteriophage $\lambda$ seems especially attractive. Phage $\lambda$ is assembled inside the cytosol of *E. coli* and released upon lysis of the cell. Thus, proteins displayed on its surface need not be transported across the cell membrane and folded in the periplasm of *E. coli*. As already mentioned, the secretion into the periplasm can be especially problematic for cytosolic proteins and is one of the major constraints of filamentous phage display.

So far display of foreign peptides and proteins on phage $\lambda$ has been accomplished by fusion to the major tail tube protein gpV (Dunn 1995; Maruyama et al. 1994) as well as to the major head protein gpD (Mikawa et al. 1996; Sternberg and Hoess 1995). Examples of proteins displayed include β-lactamase (Mikawa et al. 1996), IgG-binding domains (Mikawa et al. 1996; Sternberg and Hoess 1995) β-galactosidase (Dunn 1995; Maruyama et al. 1994; Mikawa et al. 1996), as well as peptides, peptide libraries and epitope libraries (Kuwabara et al. 1997). Based on comparison of the published data, fusion to gpD system seems to be superior to fusion to gpV for a number of reasons: (1) gpD appears as prominent protrusions on the capsid surface, making fusion peptides easily accessible (Dokland and Murialdo 1993); (2) peptides can be fused to the $NH_2$- and COOH-terminals of gpD, whereas peptides can only be fused to the COOH-terminal of gpV (Mikawa et al. 1996); (3) phage with a genome smaller than 82% of wild-type can form stable phage in the absence of gpD, indicating that gpD fusion proteins should not interfere with infectivity and phage assembly.

There are 405 copies of gpD per phage head. Using a elegant conditional chain termination scheme, which produces the foreign protein, gpD and the corresponding fusion protein, it has been shown that homo-multimeric proteins such as β-galactosidase can be displayed (Mikawa et al. 1996). The number of fusion proteins displayed per phage can be varied by using *E. coli* strains with different suppressor activities.

Although the λ display systems seems to be complementary to filamentous phage display and an attractive alternative for certain applications, it has to be kept in mind that, up to now, few examples of its application have been documented. As with filamentous phage display, opportunities and limitations will be come apparent as the system is used for broader and more demanding applications.

# 4 Phage Display of Combinatorial Libraries

## 4.1 Peptide Libraries

Peptide libraries can be displayed either multivalently, as fusion to gIIIp or gVIIIp, or monovalently, as fusion to gIIIp. Due to the chelate or avidity effect, multivalent display allows the recovery of peptides binding to target molecules with low or modest affinity. Due to their low affinity, those peptides may fail to be isolated when displayed monovalently (Wrighton et al. 1996). However, it is possible to combine the two approaches to first isolate low affinity peptides using multivalent display, and then to improve these peptides using monovalent gIIIp display (Wrighton et al. 1996).

Peptides can be displayed linearly, i.e. non-constrained, or cyclically, i.e. constrained. Constrained peptide libraries are constructed by introducing two cysteine residues flanking the random peptide sequences; following the transport of a peptide-gIIIp fusion protein into the periplasm, the two cysteine residues are oxidized to form a disulfide bond. Constraining the displayed peptides reduces the conformational flexibility substantially and thus is expected to lead to better binding affinity than in non-constrained peptides. There are very few comprehensive comparisons of linear vs cyclic peptide libraries, but even from those limited studies, constrained peptides seem to perform better (Devlin et al. 1990; Giebel et al. 1995; Katz 1995). Alternative methods for presenting constrained peptide libraries are discussed in Sect. 4.2, Protein Libraries.

## 4.2 Protein Libraries

Proteins, including antibodies, are displayed either as gVIIIp or, most frequently, as gIIIp fusion proteins. Fusion to gVIIIp is restricted to complementation by wild-type gVIIIp from helper phage (Kang et al. 1991).

Intuitively, only secreted or extracellular proteins such as antibodies or extracellular domains of surface receptors are expected to be functionally displayed. Indeed, most of the displayed proteins belong to these classes of protein families, including antibodies, cytokines such as human growth hormone (Bass et al. 1990; Lowman and Wells 1993) and interleukins (Gram et al. 1993; Clackson and Wells 1994), receptors such as PDGF receptor (Chiswell and Mccafferty 1992), and enzymes that are naturally located in the periplasm such as β-lactamase (Soumillion et al. 1994) and alkaline phosphatase (Mccafferty et al. 1992). However, experience has shown that it is difficult to predict whether a protein can be exported and folded in the periplasm of *E. coli* and even a few cytosolic proteins have been functionally displayed on phage. For instance, a class of transcription factors, zinc fingers, which are located in the cell nucleus (see Sect. 5.1, Ligand Selection) as well as cytoplasmic proteins such as ciliary neurotrophic factor can be displayed and functionally selected (Saggio et al. 1995).

The most important applications of phage display of protein libraries have been in the field of antibody selections and engineering. This can be explained by the importance of antibodies in medical and diagnostic applications as well as the fact that functional expression of antibodies is possible in the periplasm of *E. coli*. For a comprehensive review on phage display of antibody libraries, the reader is referred to two recent reviews (Griffith and Duncan 1998; Rader and Barbas 1997).

Proteins displayed on phage are also used as scaffolds for constrained peptide libraries or as alternatives to antibodies (Nord et al. 1997; Martin et al. 1996; Ku and Schultz 1995; Mcconnell and Hoess 1995; Smith et al. 1998). For example, Ku and Schultz (1995) have constructed a protein library from the four-helix bundle protein cytochrome β562 by randomizing the two loops at one end of the four-helix bundle and selected for binders against a small organic hapten. Mutants which fold into native-like conformations and bind the BSA conjugated hapten with micromolar dissociation constants were isolated. Binding affinities of the mutants are comparable to that of a monoclonal antibody which has been raised against this hapten.

The use of "knottins" as a small binding protein in phage selections has recently been published by Griffiths and coworkers (Smith et al. 1998). Knottins are small, disulfide-containing proteins that bind to a variety of targets ranging from lipids to proteins. Starting with a known knottin, seven residues, located in the native binding site of the protein, were randomized. The corresponding knottin library was selected for binding to cellulose (the natural target of wild-type) and three different proteins, including alkaline phosphatase. Indeed, specific binders for cellulose and alkaline phosphatase were isolated. Although binders to the two other proteins were not present in the initial library, knottins appear to be attractive candidates for the generation of small, easily folded proteins with binding activities. In addition, their small size makes their structure determination by NMR as well as their chemical synthesis possible.

## 4.3 cDNA Libraries

Phage gIIIp and gVIIIp display are restricted to the fusion of foreign peptides and proteins to the $NH_2$-terminus of gIIIp and gVIIIp, as the COOH-terminal domains of both proteins are embedded in the phage particle. The presence of stop codons at the 3′-end of non-translated regions of eukaryotic mRNA imposes a challenge for displaying cDNA libraries $NH_2$-terminal to gIIIp. Three methods have been reported that have been or can be used for the display of cDNA libraries on filamentous bacteriophage.

Crameri and Suter (1993) have made use of the strong association of the two leucine zipper polypeptides Jun and Fos to display cDNA products on phage. Jun was fused to the $NH_2$-terminus of gIIIp, the cDNA library was fused to the COOH-terminus of Fos. Co-expression of both fusion proteins from the same phagemid gave rise to association of Jun-gIIIp with Fos-cDNA in the periplasm. In order to prevent dissociation of the Jun-Fos complex, a disulfide bond was engineered into the leucine zipper. After infection with helper phages, phage particles displaying cDNA libraries could be produced and selected. Using this method, *Aspergillus fumigatus* cDNA clones binding to human serum IgE were isolated (Crameri et al. 1994).

Alternatively, cDNA can be fused to the COOH-terminal of the two $NH_2$-terminal domains of gIIIp, thus physically separating the $NH_2$-terminal of gIIIp from its COOH-terminus domain (see Sect. 3.1, Filamentous Bacteriophage Display). Gramatikoff et al. (1994) have fused a human B-cell cDNA library COOH-terminal to gIIIpN and selected for proteins interacting with Jun, fused to the $NH_2$-terminus of gIIIpC. Two clones predicted to interact with Jun were isolated, demonstrating the feasibility of the approach (Gramatikoff et al. 1995). In another application of this method, the interaction of HIV-1 reverse transcriptase with β-actin was shown (Hottiger et al. 1995).

The third approach made use of the finding that gVIp possesses a solvent-exposed COOH-terminal and that proteins fused to the COOH-terminal of gVIp can be functionally displayed on phage. Using this approach, Jespers et al. (1995) have selected from a *Ancylostoma caninum* cDNA library two serine protease inhibitors of trypsin and factor Xa, respectively. This system was further optimized by creating a vector that combines the high cloning efficiency of lambda vectors (Jespers et al. 1996) with gVIp phage display.

However, it has to be kept in mind that, due to the general constraints of phage display, not all proteins encoded in a cDNA library will be displayed in their native or functional form. Furthermore, proteins which require a free $NH_2$-terminus for folding or function cannot be selected using the methods mentioned here.

# 5 Applications

## 5.1 Ligand Selection

The selection of ligands from combinatorial libraries to specific targets has been the main application of phage display (Scott and Smith 1990; Devlin et al. 1990; Cwirla et al. 1990). Of particular interest has been the identification of small peptides as mimics of proteins in protein–protein interactions (Cortese et al. 1995). Experiments that demonstrate the enormous potential of phage peptide libraries in this area have included the selection of peptides that bind to the erythropoietin (EPO) receptor (Wrighton et al. 1996). To allow for the isolation of low-affinity binders to the receptor, initial selections were performed with multivalent peptide libraries fused to gVIIIp. Selected peptides were further optimized using the monovalent gIIIp display systems, yielding a cyclic peptide of 20 amino acids which binds the EPO receptor with nanomolar affinity. Remarkably, the peptide acts as an agonist of EPO and can stimulate erythropoiesis in mice, a result supported by the crystal structure of the peptide-receptor complex (Livnah et al. 1996). The structure revealed that the selected peptide forms a homodimer which induces dimerization of the EPO receptor, an event that is essential for its physiological activity. As the stabilization of the dimer should enhance its biological activity, the peptide was chemically cross-linked (Wrighton et al. 1997), resulting in an EPO mimetic of 20 amino acids with an affinity of 2nM, only two orders of magnitude below the affinity of the 165-amino acid cytokine EPO itself. Furthermore, the dimer also showed increased potency in cell-based assays and in mice. These and other experiments (Yanofsky et al. 1996) clearly demonstrate the potential of combinatorial peptide libraries for the functional mimicry of large proteins by small, unrelated peptides. Furthermore, the capability to rapidly identify peptide leads for the manipulation of protein–protein interactions has applications for drug discovery and other pharmaceutical applications.

Phage display has also been successfully used to study and redesign protein-DNA interactions (Rebar and Pabo 1994; Jamieson et al. 1994; Choo and Klug 1994, Wu et al. 1995). Zinc fingers have been selected based on their ability to bind oligonuleotides; randomization of the residues in the zinc fingers that mediate DNA binding and selection against various oligonucleotides led to zinc fingers with altered specificities. A general method that allows the in vitro selection of zinc finger proteins for diverse DNA target sites has recently been published (Greisman and Pabo 1997). The method relies on a stepwise approach, gradually extending a zinc finger protein across the desired target sequence, adding and optimizing one finger at a time. As a guarantee that the starting zinc finger protein binds with sufficient affinity to the chosen target site, the first randomized finger is fused to two wild-type zinc fingers. Accordingly, the DNA used in the selections has a wild-type binding site next to the target site. After the selection of the first finger, one of the wild-type fingers is discarded and the second randomized finger is added and optimized. After the selection of the second finger, the last wild-type finger is removed and the third zinc finger is optimized against the target site. Using this stepwise procedure, zinc fingers for a TATA box, a p53 site and a nuclear receptor element were obtained. Binding constants and specificities of these artificial DNA-binders were comparable to wild-type zinc fingers, opening the door for possible therapeutic applications.

An intriguing application of phage display with pharmaceutical and medical implications has been highlighted by the in vivo selection of peptides for organ-selective targeting (Pasqualini and Ruoslahti 1996). Here, cyclic peptide libraries displayed on phage have been injected intravenously into mice and subsequently recovered from the brain and the kidney. After three rounds of in vivo selections, different peptide motifs were identified which were capable of selectively targeting the corresponding phage to brain or kidney blood vessels. One of the identified peptides was synthesized and was shown to be capable of inhibiting the targeting of the corresponding phage to the brain. Another advantage of this procedure is that it not only selects for peptides with specific binding to markers on endothelial surfaces, but also for stability under the selection conditions (the peptide has to reach its target inside the bloodstream before being degraded).

## 5.2 Protein Folding and Design

Combinatorial approaches offer tremendous possibilities in the area of protein folding and design. Since the first publications in 1995 (Gu et al. 1995, O'Neil et al. 1995), there have been also an increasing number of exciting examples in which phage display has been used to study problems in this area. Generally, in these experiments one of the properties of the folded state of the displayed protein, such as binding to an immobilized target, is subject to the selection process. For example, phage display was used to study the role of turns in the folding of the B1 IgG binding domain of peptostreptococcal protein L (referred to as protein L; Gu et al. 1997). The structure of protein L (62 amino acids) consists of a four-stranded β-sheet packed against an α-helix, the order of secondary structure elements is ββαββ. To obtain insight into the role of the two β-turns in the folding of the protein, both turns were independently randomized and folded mutants selected from phage libraries based on their ability to bind IgG. Characterization of selected mutants revealed a significantly higher sequence variability in one of the two turns. Based on the sequence data and biophysical properties of selected mutants, the authors concluded that the role of β-turns in proteins in general is strongly context-dependent, and that in protein L only one of the two turns is formed in the transitions state of the folding process.

The design of miniaturized proteins might prove to be valuable for understanding pivotal determinants of protein folding, as illustrated by an approach taken by Braisted and Wells (1996), who systematically minimized the Z-domain of protein A. The Z-domain is a three-helix bundle of 59 amino acids which binds tightly to IgG1. The binding contacts are mediated through the first two helices, whereas the third helix is only necessary for stabilizing the overall fold. In order to reduce the three helix motive of the Z-domain to a two helix motive which retains binding to IgG1, Braisted and Wells (1996) used a stepwise, combinatorial strategy. First, the third helix was deleted, residues in the truncated version of the Z-domain that previously contacted the third helix were randomized and the resulting protein selected against the immobilized receptor. The selected consensus peptide was then randomized at the hydrophobic core between the remaining two helices. Affinity selections again yielded a consensus peptide, which was mutagenized at the positions that mediate contact to IgG. After the final affinity selections, the researchers isolated a size-minimized protein Z domain that bound to IgG with an affinity comparable to wild-type, despite deleting 26 of the 59 amino acids of the wild-type protein. In a subsequent report (Starovasnik et al. 1997), the solution structure of the minimized domain was determined. Based on this structural information, a disulfide bridge was introduced, resulting in a much more stable variant with a nine-fold improvement in affinity.

A study of protein–protein interfaces was recently conducted by researchers at Genentech and at the University of California, San Francisco (Atwell et al. 1997a). The starting point was the high-affinity interface between human growth hormone (hGH) and the extracellular domain of its receptor (hGHbp). To study how a functionally disruptive mutation in one binding partner can be complemented by mutations in the other binding partner, a critical tryptophan residue was mutated to an alanine (W104A), creating a large cavity in the hGHbp. The binding partner hGH was functionally displayed on phage and five residues that contact the deleted tryptophan residue were randomized and the corresponding library was selected for binding to the mutated hGHbp. The most frequently selected clone contained four mutations in the randomized positions and, remarkably, one mutation that was not introduced by random mutagenesis. To understand how the W104A mutation is complemented by the selected mutations, a crystal structure of the mutant complex was obtained. The large cavity formed by the W104A mutation is largely filled by a combination of the mutations in hGH and conformational changes in both binding partners. Remarkably, the local structural changes near the site of the W104A mutation result in substantial global changes that affect the entire interface: Groups as far as 15Å away from the W104A mutation move up to 3Å relative to their position in the wild-type complex. The authors suggest that this observed structural plasticity might be a means for protein–protein interfaces to adapt to mutations as they co-evolve.

A similar "knobs-into-holes" approach was used to select for combinations of interface residues of antibody $C_H3$ domains that preferentially form heterodimers instead of homodimers (Atwell et al. 1997b). A knob mutant of one $C_H3$ domain was created by changing a threonine at the domain interface to a tryptophan and this mutant was fused to a peptide flag. A library of hole mutants of a $C_H3$ domain was then created by randomizing residues in proximity to the knob of the partner domain and fused to gIIIp. Using an anti-flag antibody, coexpression of the two constructs from a common phagemid allows selection of phage displaying stable heterodimers. Demonstrating the power of combinatorial methods, the selected heterodimers were more stable than heterodimers previously generated by rational design.

Disulfide bonds in naturally occurring antibodies are crucial for their correct folding and stability. However, phage display can also be used to select scFv antibodies without the conserved disulfide bonds (PROBA et al. 1998). By using DNA shuffling and phage display, disulfide-free scFvs were isolated with thermodynamic stabilities comparable to wild-type. The functional expression of antibody fragments under the reducing conditions of the cytoplasm opens the door for many possible therapeutic and diagnostic applications (BIOCA and CATTANEO 1995).

Clearly, regardless of how ingenious the employed selection scheme is, it has to be kept in mind that phage selections using protein libraries always pick out a combination of properties of the fusion protein, including expression, stability, folding and activity. This fact has been highlighted recently by a number of model selections by Plückthun and coworkers (PEDRAZZI et al. 1997). A model library was created by mutating 11 residues of a fluorescein binding scFv antibody to alanine. This model library was used in three subsequent rounds of selections for binding to fluorescein. Sequencing of 44 selected clones revealed that a mutant no longer carrying an exposed tryptophan residue was preferentially selected, although it binds the antigen with affinities comparable to a number of other mutants present in the model library at the beginning of the selection. This finding stresses the important fact that not only affinity, but also parameters such as folding and stability, play a significant role in phage selections.

## 5.3 Enzyme Design

Phage display of proteins and enzymes (see also Sect. 4.3, Display of Protein Libraries) has also opened the door for novel approaches in the area of enzyme design. In particular, the selection of novel catalytic antibodies using phage display is becoming increasingly important as an alternative or complement to traditional methods. In general, generation of catalytic antibodies relies on the diversity of the immune system to generate binders to almost any foreign molecule as well as on mechanistic insights in catalysis (SCHULTZ and LERNER 1995). Although impressive rate enhancements have been achieved in a number of cases, catalytic antibodies usually do not rival the performance of natural enzymes. A main reason for this can be attributed to the fact that these antibodies are selected for binding to a transition state analog and not for catalysis directly. Recently, it has been demonstrated that phage display offers the opportunity to create a link between antibody catalysis of a particular reaction and the replication of the corresponding phages displaying the catalytic antibody. In general, these selection schemes lead to a physical change of the displayed antibody, which allows its isolation and subsequent amplification out of a pool of unreactive molecules. The first example of this approach has been the selection of antibodies possessing reactive cysteines using immobilized, activated disulfides. Incubation of semisynthetic Fab antibody phage libraries with the activated disulfide leads to selective immobilization of reactive antibodies due to a disulfide exchange reaction between the immobilized disulfide and a reactive cysteine of the displayed antibody. These experiments resulted in the selection of antibodies that were mechanistically analogous to known thiol proteases, which catalyze the hydrolysis of thioesters (Janda et al. 1994). A much more general scheme, which directly couples catalysis and phage replication, has recently been reported (Janda et al. 1997). This approach relies on the careful design of an immobilized, modified substrate, which, upon turnover by the catalytic antibody, creates a reactive species that covalently attaches to the antibody, leading to immobilization of catalytic clones. The remarkably high catalytic activity of selected antibodies (approximately $10^5$-fold rate enhancement) demonstrates the potential of this approach. Furthermore, using a slightly different strategy, the in vitro selection of antibodies from phage libraries that catalyze the hydrolysis of unactivated amide bonds, a reaction that is extremely difficult to catalyze, has also been reported recently (Gao et al. 1998).

An approach which uses the technique of selectively infective phage (see Sect. 3.1, "Filamentous Bacteriophage Display Systems") to create a link between catalysis and replication has also been introduced (Gao et al. 1997). In a model selection, a catalytic antibody operating by covalent catalysis and, under appropriate conditions, capable of forming a stable covalent intermediate with its substrate, was fused to gIIIpC. Accordingly, its substrate was chemically fused to gIIIpN. As a result, addition of the reactive antibody to the substrate fused to gIIIpN restores infectivity, making the formation of the covalent intermediate a selectable process.

Phage display of a randomized catalytic antibody can also be used to perform in vitro affinity maturation of catalytic antibodies to optimize the binding to transition state analogs (Baca et al. 1997; Fujii et al. 1998).

Obviously, the use of phage display is not limited to the selection of catalytic antibodies, it can also be used for the selection of novel enzymatic activities starting from known enzymes or for the improvement of existing enzyme activity. Furthermore, phage display can be a valuable tool for studying enzyme mechanisms (Hansson et al. 1997).

## 6 Conclusions

Since the emergence of the first examples of its use in 1990, phage display of combinatorial peptide and protein libraries has become an indispensable technique in various laboratories. Major applications will continue to be the selection of peptides as lead compounds for pharmaceutical applications and of antibodies for medical and diagnostic applications. Furthermore, the screening of cDNA libraries as well as approaches such as protein–protein interaction mapping (library against library screening) will become increasingly important. Here, the recent development of techniques such as selectively infective phage hold great promise for further applications. However, the screening of cDNA libraries also highlights limitations of phage display: functional display of proteins on filamentous phage requires compatibility with transport to and folding in the periplasm of *E. coli*, resulting in inevitable preselections. In these areas phage λ display or other techniques are expected to complement traditional filamentous phage display. Another severe limitation of using phage display for applications such as protein–protein interaction mapping is the size of the library that can be used, which is limited to approximately $10^9$. Since comprehensive protein–protein interaction mapping involves at least two libraries, the size of each library cannot exceed $10^5$, too small to cover most of the available cDNA or genomic libraries. Clearly, progress in this area will depend on the developments of new techniques and methodologies for phage display of combinatorial protein libraries.

Generally, the successful use of phage display for a particular application requires a careful choice of the display format and the selection scheme as well as an appreciation of the biology of filamentous phage and of the properties of the displayed protein.

*Acknowledgements.* The authors would like to thank Robert Damoiseaux for critical reading of the manuscript.

## References

Atwell S, Ultsch M, De Voss AM, Wells JA (1997a) Structural plasticity in a remodeled protein–protein interface. Science 278:1125–1128

Atwell S, Ridgway JBB, Wells JA, Carter P (1997b) Stable heterodimers from remodeling the domain interface of a homodimer using a phage display library. J Mol Biol 270:26–35

Baca M, Scanlan TS, Stephenson RC, Wells JA (1997) Phage Display of a catalytic antibody to optimize affinity for transition state binding. Proc Natl Acad Sci USA 94:10063–10068

Barbas III CF, Kang AS, Lerner RA, Benkovic SJ (1991) Assembly of combinatorial antibody libraries on phage surface: the gene III site. Proc Natl Acad Sci USA 88:7978–7982

Bass S, Greene R, Wells JA (1990) Hormone phage: an enrichment method for variant proteins with altered binding properties. Proteins 8:309–314

Bass SH, Mulkerrin MG, Wells JA (1991) A systematic mutational analysis of hormone-binding determinants in the human growth hormone receptor. Proc Natl Acad Sci USA 88:4498–4502

Biocca S, Cattaneo A (1995) Intracellular immunization: antibody targeting to subcellular compartments. Trends Cell Biol 5:59–63

Boder ET, Wittrup KD (1997) Yeast surface display for screening combinatorial polypeptide libraries. Nature Biotechnol 15:553–557

Bothmann H, Plückthun A (1998) Selection for a periplasmic factor improving phage display and functional periplasmic expression. Nature Biotechnol 16:376–380

Boublik Y, Di Bonito P, Jones IM (1995) Eukaryotic virus display: Engineering the major surface glycoprotein of the Autographa californica nuclear polyhedrosis virus (AcNPV) for the presentation of foreign proteins on the virus surface. BioTechnology 13:1079–1084

Braisted A, Wells JA (1996) Minimizing a binding domain from protein A. Proc Natl Acad Sci USA 93:5688–5692

Chen R, Henning U (1996) A periplasmic protein (Skp) of *Escherichia coli* selectively binds a class of outer membrane proteins. Mol Microbiol 19:1287–1294

Chiswell DJ and McCafferty J (1992) Phage antibodies: will new 'coliclonal' antibodies replace monoclonal antibodies? Trends in Biotechnol 10:80–84

Choo Y, Klug A (1994) Selection of DNA binding sites for zinc fingers using rationally randomized DNA reveals coded interactions. Proc Natl Acad Sci USA 91:11163–11167

Clackson T, Wells JA (1994) In vitro selection from protein and peptide libraries. Trends Biotechnol 12:173–184

Cortese R, Monaci P, Nicosia A, Luzzago A, Felici F, Galfré G, Pessi A, Tramontao A, Sollazzo M (1995) Identification of biologically active peptides using random libraries displayed on phage. Curr Opin Biotechnol 6:73–80

Crameri R, Suter M (1993) Display of biologically active proteins on the surface of filamentous phages: a cDNA cloning system for selection of functional gene products linked to the genetic information responsible for their production. Gene 137:69–75

Crameri R, Jaussi R, Menz G, Blaser K (1994) Display of expression products of cDNA libraries on phage surface: a versatile screening system for selective isolation of genes by specific gene-product/ligand interaction. Eur J Biochem 226:53–58

Cwirla SE, Peters EA, Barrett RW, Dower WJ (1990) Peptides on phage: a vast library of peptides for identifying ligands. Proc Natl Acad Sci USA 87:6378–6382

Devlin JJ, Panganiban LC, Devlin PE (1990) Random peptide libraries: a source of specific protein binding molecules. Science 249:404–406

Dokland T, Murialdo H (1993) Structural transitions during maturation of bacteriophage lambda capsids. J Mol Biol 233:682–694

Duenas M, Borrebaeck CAK (1994) Clonal selection and amplification of phage displayed antibodies by linking antigen recognition and phage replication. Bio/Technology 12:999–1002

Dunn IS (1995) Assembly of functional bacteriophage lambda virions incorporating COOH-terminal peptide or protein fusions with the major tail protein. J Mol Biol 248:497–506

Dunn IS (1996) Phage display of proteins. Curr Opin Biotechnol 7:547–553

Endemann H, Model P (1995) Location of filamentous phage minor coat proteins in phage and in infected cells. J Mol Biol 250:496–506

Efimov VP, Nepluev IV, Mesyanzhinov VV (1995) Bacteriophage T4 as a surface display vector. Virus Genes 10:173–177

Fujii I, Fukuyama S, Iwabuchi Y, Tanimura R (1998) Evolving catalytic antibodies in a phage-displayed combinatorial library. Nature Biotechnol 16:463–467

Gao C, Lavey BJ, Lo C-HL, Datta A, Wentworth Jr. P, Janda KD (1998) Direct selection for catalysis from combinatorial antibody libraries using a boronic acid probe: Primary amide bond hydrolysis. J Am Chem Soc 120:2211–2217

Gao C, Lin C-H, Lo C-HL, Mao S, Wirsching P, Lerner RA, Janda KD (1997) Making chemistry selectable by linking it to infectivity. Proc Natl Acad USA 94: 11777–11782

Garrard LJ, Yang M, O'Connell MP, Kelley RF, Henner DJ (1991) Fab assembly and enrichment in a monovalent phage display system. Bio/Technology 9:1373–1377

Giebel LB, Cass RT, Milligan DL, Young DC, Arze R, Johnson CR (1995) Screening of cyclic peptide phage libraries identifies ligands that bind streptavidin with high affinities. Biochemistry 34:15430–15435

Goldsmith ME, Konigsberg WH (1977) Adsorption protein of bacteriophage fd: Isolation, molecular properties, and location in virus. Biochemistry 16:2686–2694

Gram H, Strittmatter U, Lorenz M, Gluck D, Zenke G (1993) Phage display as a rapid gene expression system: Production of bioactive cytokine-phage and generation of neutralizing monoclonal antibodies. J Immunol Methods 161:169–176

Gramatikoff K, Georgiev O, Schaffner W (1994) Direct interaction rescue, a novel filamentous phage technique to study protein–protein interactions. Nuc Acids Res 22:5761–5762

Gramatikoff K, Schaffner W, Georgiev O (1995) The leucine zipper of c-Jun binds to ribosomal protein L18a: a role in Jun protein regulation? Biol Chem Hoppe-Seyler 376:321–325

Grant RA, Lin TC, Konigsberg W, Webster RE (1981) Structure of the filamentous bacteriophage f1: Location of the A, C and D minor coat proteins. J Biol Chem 256:539–546

Gray CW, Brown RS, Marvin DA (1981) Adsorption complex of filamentous fd virus. J Mol Biol 146:621–627

Greisman HA, Pabo CO (1997) A general strategy for selecting high-affinity zinc finger proteins for diverse DNA target sites. Science 275:657–661
Griffith AD, Duncan AR (1998) Strategies for selection of antibodies by phage display. Curr Opin Biotechnol 9:102–108
Gu H, Kim D, Baker D (1997) Contrasting roles for symmetrically disposed β-turns in the folding of a small protein. J Mol Biol 274:588–596
Gu H, Yi Q, Bray ST, Riddle DS, Shiau AK, Baker D (1995) A phage display system for studying the sequence determinants of protein folding. Protein Sci 4:1108–1117
Hansson LO, Widersten M, Mannervik B (1997) Mechanism-based phage display selection of active-site mutants of human glutathione transferase A1–1 catalyzing $S_NAr$ reactions. Biochemistry 36:11252–11260
Holliger P, Riechmann L (1997) A conserved infection pathway for filamentous bacteriophages is suggested by the structure of the membrane penetration domain of the minor coat protein g3p from phage fd. Structure 5:265–275
Hottiger M, Gramatikoff K, Georgiev O, Chaponnier C, Schaffner W, Hubscher U (1995) The large subunit of HIV-1 reverse transcriptase interacts with β-actin. Nuc Acids Res 23:736–741
Iannolo G, Minenkova O, Petruzzelli R, Cesareni G (1995) Modifying filamentous phage capside: Limits in the size of the major capsid protein. J Mol Biol 248:835–844
Jacobson A (1972) Role of F pili in the penetration of bacteriophage f1. J Virol 10:835–843
Jamieson AC, Kim S-H, Wells JA (1994) In vitro selection of zinc fingers with altered DNA-binding specificity. Biochemistry 33:5689–5695
Janda KD, Lo C-HL, Li T, Barbas CF III, Wirsching P, Lerner RA (1994) Direct selection for a catalytic mechanism from combinatorial antibody libraries. Proc Natl Acad Sci USA 91:2532–2536
Janda KD, Lo L-C, Lo C-HL, Sim M-M, Wang R, Wong C-H, Lerner RA (1997) Chemical selection for catalysis in combinatorial antibody libraries. Science 275:945–948
Jiang J, Abu-Shilbayeh L, Rao VB (1997) Display of a PorA peptide from Neisseria meningitidis on the bacteriophage T4 capsid surface. Infect Immun 65:4770–4777
Jespers LS, De Keyser A, Stanssens PE (1996) LambdaZLG6: a phage lambda vector for high-efficiency cloning and surface expression of cDNA libraries on filamentous phage. Gene 173:179–181
Jespers LS, Messens JH, De Keyser A, Eeckhout D, Van Den Brande I, Gansemans YG, Lauwereys MJ, Vlasuk GP, Stanssens PE (1995) Surface expression and ligand-based selection of cDNAs fused to filamentous phage gene VI. Bio/Technology 13:378–382
Kang AS, Barbas III CF, Janda KD, Benkovic SJ, Lerner RA (1991) Linkage of recognition and replication functions by assembling combinatorial antibody Fab libraries along phage surface. Proc Natl Acad Sci USA 88:4363–4366
Katz BA (1995) Binding to protein targets of peptidic leads discovered by phage display: crystal structures of streptavidin-bound linear and cyclic peptide ligands containing the HPQ sequence. Biochemistry 34:15421–15429
Knappik A, Krebber C, Plückthun A (1993) The effect of folding catalysts on the in vivo folding process of different antibody fragments expressed in *Escherichia coli*. Biotechnology 11:77–83
Krebber C, Spada S, Desplancq, Krebber A, Plückthun A (1995) Co-selection of cognate antibody-antigen pairs by selectively-infective phages. FEBS Letters 377:227–231
Krebber C, Spada S, Desplancq, Krebber A, Ge L, Plückthun A (1997) Selectively-infective phage (SIP): A mechanistic dissection of a novel in vivo selection for protein-ligand interactions. J Mol Biol 268:607–618
Ku J, Schultz PG (1996) Alternate protein frameworks for molecular recognition. Proc Natl Acad Sci USA 92:6552–6556
Kuwabara I, Maruyama H, Mikawa YG, Zuberi RI, Liu FT, Maruyama IN (1997) Efficient epitope mapping by bacteriophage l surface display. Nature Biotechnol 15:74–78
Lin TC, Webster RE, Konigsberg W (1980) Isolation and characterization of the C and D proteins coded by gene IX and gene VI in the filamentous bacteriophage f1 and fd. J Biol Chem 255:10331–10337
Lindqvist BH, Naderi S (1995) Peptide presentation by bacteriophage P4. FEMS Microbiol Rev 17:33–39
Livnah O, Stura E, Johnson DL, Middleton SA, Mulcahy LS, Wrighton NC, Dower WJ, Jolliffe LK, Wilson IA (1996) Functional mimicry of a protein hormone by a peptide agonist: The EPO receptor complex at 2.8 Å. Science 273:464–471
Lopez J, Webster RE (1983) Morphogenesis of filamentous bacteriophage f1: Orientation of extrusion and production of polyphage. Virology 127:177–193
Lowman HB, Wells JA (1993) Affinity maturation of human growth hormone by monovalent phage display. J Mol Biol 234:564–578

Lubkowski J, Hennecke F, Plückthun A, Wlodawer A (1998) The structural basis of phage display elucidated by the crystal structure of the N-terminal domains of g3p. Nature Struct Biol 5:140–147
Malik P, Terry TD, Gowda LR, Langara A, Petukhov SA, Symmons MF, Welsh LC, Marvin DA, Perham RN (1996) Role of capsid structure and membrane protein processing in determining the size and copy number of peptides displayed on the major coat protein of filamentous bacteriophage. J Mol Biol 260:9–21
Malmborg A-C, Söderlind E, Frost L, Borrebaeck CAK (1997) Selective phage infection mediated by epitope expression on F pilus. J Mol Biol 273:544–551
Markland W, Roberts BL, Saxena MJ, Guterman SK, Ladner RC (1991) Design, construction and function of a multicopy display vector using fusions to the major coat protein of bacteriophage M13. Gene 109:13–19
Martin F, Toniatti C, Salvati AL, Ciliberto G, Cortese R, Sollazzo M (1996) Coupling protein design and in vitro selection strategies: improving specificity and affinity of a designed beta-protein IL-6 antagonist. J Mol Biol 255:86–97
Maruyama IN, Maruyama HI, Brenner S (1994) λfoo: A λ phage vector for the expression of foreign proteins. Proc Natl Acad Sci USA 91:8273–8277
Mikawa YG, Maruyama IN, Brenner S (1996) Surface display of proteins on bacteriophage λ heads. J Mol Biol 262:21–30
McCafferty J, Jackson RH, Chiswell DJ. (1992) Phage-enzymes: expression and affinity chromatography of functional alkaline phosphatase on the surface of bacteriophage. Protein Eng 4:955–961
McConnell SJ, Hoess RH (1995) Tendamistat as a scaffold for conformationally constrained phage peptide libraries. J Mol Biol 250:460–470
Model P, Russel M (1988) Filamentous Bacteriophage. In: Bacteriophages 2:375–456. Plenum. New York
Nord K, Gunneriusson E, Ringdahl J, Stahl S, Uhlen M, Nygren PA (1997) Binding proteins selected from combinatorial libraries of an alpha-helical bacterial receptor domain. Nature Biotechnol 15:772–777
Ohkawa I, Webster RE (1981) The orientation of the major coat protein of bacteriophage f1 in the cytoplasmic membrane of *Escherichia coli*. J Biol Chem 256:9951–9958
O'Neil KT, Hoess RH, Raleigh DP, DeGrado WP (1995) Thermodynamic genetics of the folding of the B1 immunoglobulin-binding domain from streptococcal protein G. Proteins Struct Funct Genet 21:11–21
Pasqualini R, Ruoslahti E (1996) Organ targeting in vivo using phage display peptide libraries. Nature 380:364–366
Pedrazzi G, Schwesinger F, Honegger A, Krebber C, Plückthun A (1997) Affinity and folding properties both influence the selection of antibodies with the selectively infective phage (SIP) methodology. FEBS Lett 415:289–293
Proba K, Wörn A, Honegger A, Plückthun A (1998) Antibody scFv fragments without disulfide bonds made by molecular evolution. J Mol Biol 275:245–253
Rakonjac J, Model P (1994) The influence of the reducing agent dithiothreitol In vitro on infectivity of f1 particles. Gene 142:153–154
Rader C, Barbas III CF (1997) Phage display of combinatorial antibody libraries. Curr Opin Biotechnol 8:503–508
Rebar EJ, Pabo CO (1994) Zinc finger phage: affinity selection of fingers with new DNA-binding specificities. Science 263:671–673
Ren ZJ, Lewis GK, Wingfield PT, Locke EG, Steven AC, Black LW (1996) Phage display of intact domains at high copy number: a system based on SOC, the small outer capsid protein of bacteriophage T4. Protein Sci 5:1833–1843
Riechmann L, Holliger P (1997) The COOH-terminal domain of TolA is the coreceptor for filamentous phage infection of *E. coli*. Cell 90:351–360
Saggio I, Gloaguen I, Laufer R (1995) Functional phage display of ciliary neurotrophic factor. Gene 152:35–39
Schultz PG, Lerner RA (1995) From molecular diversity to catalysis: Lessons from the immune system. Science 269:1835–1842
Scott JK, Smith GP (1990) Searching for peptide ligands with an epitope library. Science 249:386–390
Smilowitz H (1974) Bacteriophage f1 infection: Fate of the parental coat protein. J Virol 13:94–99
Smith GP (1985) Filamentous fusion phage: Novel expression vectors that display cloned antigens on the virion surface. Science 228:1315–1317
Smith GP, Patel SU, Windass JD, Thornton JM, Winter G, Griffiths AD (1998) Small binding proteins selected from a combinatorial repertoire of knottins displayed on phage. J Mol Biol 277:317–332

Soumillion P, Jespers L, Bouchet M, Marchand-Brynaert J, Winter G, Fastrez J (1994) Selection of β-lactamase on filamentous bacteriophage by catalytic activity. J Mol Biol 237:415–422
Stahl S, Uhlen M (1997) Bacterial surface display: trends and progress. Trends Biotechnol 15:185–192
Starovasnik MA, Braistd AC, Wells JA (1997) Structural mimicry of a native protein by a minimized binding domain. Proc Natl Acad Sci USA 94:10080–10085
Sternberg N, Hoess RH (1995) Display of peptides and proteins on the surface of bacteriophage λ. Proc Natl Acad Sci USA 92:1609–1613
Wickner W (1975) Asymmetric orientation of a phage coat protein in cytoplasmic membrane of *Escherichia coli.* Proc Natl Acad Sci USA 72:4749–4753
Wrighton NC, Balasubramanian P, Barbone FP, Kashyap AK, Farrell FX, Jolliffe LK, Barrett RW, Dower WJ (1997) Increased potency of an erythropoietin peptide mimetic through covalent dimerization. Nature Biotechnol 15:1261–1265
Wrighton NC, Farrell FX, Chang R, Kashyap AK, Barbone FP, Mulcahy LS, Johnson DL, Barrett RW, Jolliffe LK, Dower WJ (1996) Small peptides as potent mimetics of the protein hormone erythropoietin. Science 273:458–463
Wu H, Yang W-P, Barbas III CF (1995) Building zinc fingers by selection: Toward a therapeutic application. Proc Natl Acad Sci USA 92:344–348
Yanofsky SD, Baldwin DN, Butler JH, Holden FR, Jacobs JW, Balasubramanian P, Chinn JP, Cwirla SE, Peters-Bhatt E, Whitehorn EA, Tate EH, Akeson A, Bowlin TL, Dower WJ, Barrett RW (1996) High affinity type I interleukin 1 receptor antagonists discovered by screening recombinant peptide libraries. Proc Natl Acad Sci USA 93:7381–7386

# In Vitro Selection Methods for Screening of Peptide and Protein Libraries

J. HANES and A. PLÜCKTHUN



## 1 Introduction

Protein ligand interactions form the basis of almost all cellular functions. The identification and improvement of the relevant ligand or receptor is therefore a focus of much of current biochemical research and the prerequisite for most pharmaceutical applications. Despite great progress, computational methods have generally not produced the accuracy required for "designing" mutations which improve a function such as binding or stability. Over the last few years, however, enormous progress in molecular biology has made it possible to imitate nature's strategy to solve the problem, the strategy of evolution. Evolution is a continuous alternation between mutation and selection. In order to apply this strategy in the laboratory over many generations both components, mutation and selection, have to be easy to perform, robust to execute and powerful in order to succeed. In this

Biochemisches Institut, Universität Zürich, Winterthurerstr. 190, CH-8057 Zürich, Switzerland chapter we will summarize the state of the art in carrying out both diversification and selection entirely in vitro, making use of cell-free translation.

All evolutionary methods must couple phenotype and genotype. Normally the carrier of the phenotype, proteins or peptides are undoubtedly the most versatile class of compounds and nature's choice in performing almost all tasks, except information storage, because of their modularity and chemical versatility. Almost all of the selection methods developed so far for peptides and proteins have either used living cells either directly, or indirectly by production of bacteriophages (Phizicky and Fields 1995). Two popular technologies illustrate the cell-based approach: the two-hybrid system (Fields and Sternglanz 1994) and phage display (Rader and Barbas 1997).

These in vivo approaches are all limited by transformation efficiency (Dower and Cwirla 1992). It is quite laborious to produce libraries of $10^{10}$ members (Vaughan et al. 1996), but enough of sequence space must be sampled to find reasonable "hits" or starting points for evolution, and such library sizes are required for undertaking difficult evolutionary tasks. Moreover, in an evolutionary approach, sequence diversification would normally take place in vitro, and thus ligation and transformation have to be repeated for every generation. Consequently, with a few exceptions (Yang et al. 1995; Low et al. 1996; Crameri et al. 1996; Moore and Arnold 1996), real evolutionary approaches with several generations have rarely been reported using in vivo methods.

In this chapter, we will summarize the state of the art for carrying out selection and diversification entirely in vitro, not using cells at any step and thereby circumventing the limitations of the in vivo methods summarized above. Libraries with more than $10^{12}$ members can now rapidly be prepared and evolved with these methods.

## 2 History of In Vitro Based Selection Methods

Already in the 1970s a series of studies showed that specific mRNA could be enriched by immunoprecipitation of polysomes with antibodies directed against the protein product (Schechter 1973; Payvar and Schimke 1979; Kraus and Rosenberg 1982). Before the general advent of molecular cloning, this was an important means of enriching the mRNA for a particular protein. Kawasaki (1991) suggested exploiting this observation as a method to enrich peptides and proteins from libraries, yet without publishing any experimental data. The idea was put into practice for the first time 3 years later by Mattheakis et al. (1994), who reported an affinity selection of short peptides from a library by using polysomes from an *E. coli* system (1994), and later by Gersuk et al. (1997) by using a wheat germ system. Significant modifications and optimizations were necessary, however, to make this concept applicable to the selection of whole, folded proteins, such as single-chain Fv (scFv) fragments of antibodies (Hanes and Plückthun 1997). Subsequently, ribosome display also was used in a eukaryotic cell-free system

(He and Taussig 1997). In a variation on this concept, it was reported that peptides could be attached to their encoding mRNAs after in vitro translation through a puromycin derivative synthetically coupled to mRNA (Nemoto et al. 1997; Roberts and Szostak 1997).

## 3 Two In Vitro Selection Schemes

The in vitro selection schemes can be divided into two groups: in the first, the polypeptide remains linked to the mRNA on the ribosome (Fig. 1). In this review, this method is referred to as "ribosome display", although other names such as polysome display (Mattheakis et al. 1994), polysome selection (Gersuk et al. 1997) or ARM selection (He and Taussig 1997) have been used. In the second strategy, after in vitro translation of the mRNA, which has to be modified to carry a puromycin derivative at the end, a covalent RNA-peptide fusion is generated by the reaction of this puromycin derivative (Fig. 2). In this review, this technique is referred to as "RNA-peptide fusion" (Roberts and Szostak 1997), although the name 'in vitro virus" has been also used (Nemoto et al. 1997). For both strategies ribosomes and all other necessary components, especially initiation and elongation factors and a specially designed mRNA, are used for in vitro translation. In ribosome display, this is performed in such a way that neither the mRNA nor its encoded peptide leave the ribosome during the ligand binding reaction; similarly, in the RNA-peptide fusion, the mRNA and the encoded peptide have to stay on the ribosome long enough for the chemical reaction to occur.

## 4 Ribosome Display

Ribosome display has been successfully performed using: (1) an *E. coli* S-30 system for display and selection of a peptides library (Mattheakis et al. 1994) or of a library of folded proteins (Hanes and Plückthun 1997; Hanes et al. 1998), (2) a wheat germ system for display and selection of a peptide library (Gersuk et al. 1997) and (3) a rabbit reticulocyte system for display and selection of folded proteins (He and Taussig 1997).

### 4.1 How Ribosome Display Works

The principle of ribosome display is shown in Fig. 1A. A DNA library, encoding a polypeptide in a special ribosome display cassette (discussed below), is either directly used for coupled in vitro transcription-translation, or first transcribed in vitro to mRNA, which is purified and used for the in vitro translation. This results in formation of ribosomal complexes (mRNA-ribosome-polypeptide), which are used for affinity selection. After removal of non-specifically bound complexes by intensive washing, RNA is isolated and used for reverse transcription and PCR. RNA can be isolated from bound ribosomal complexes either

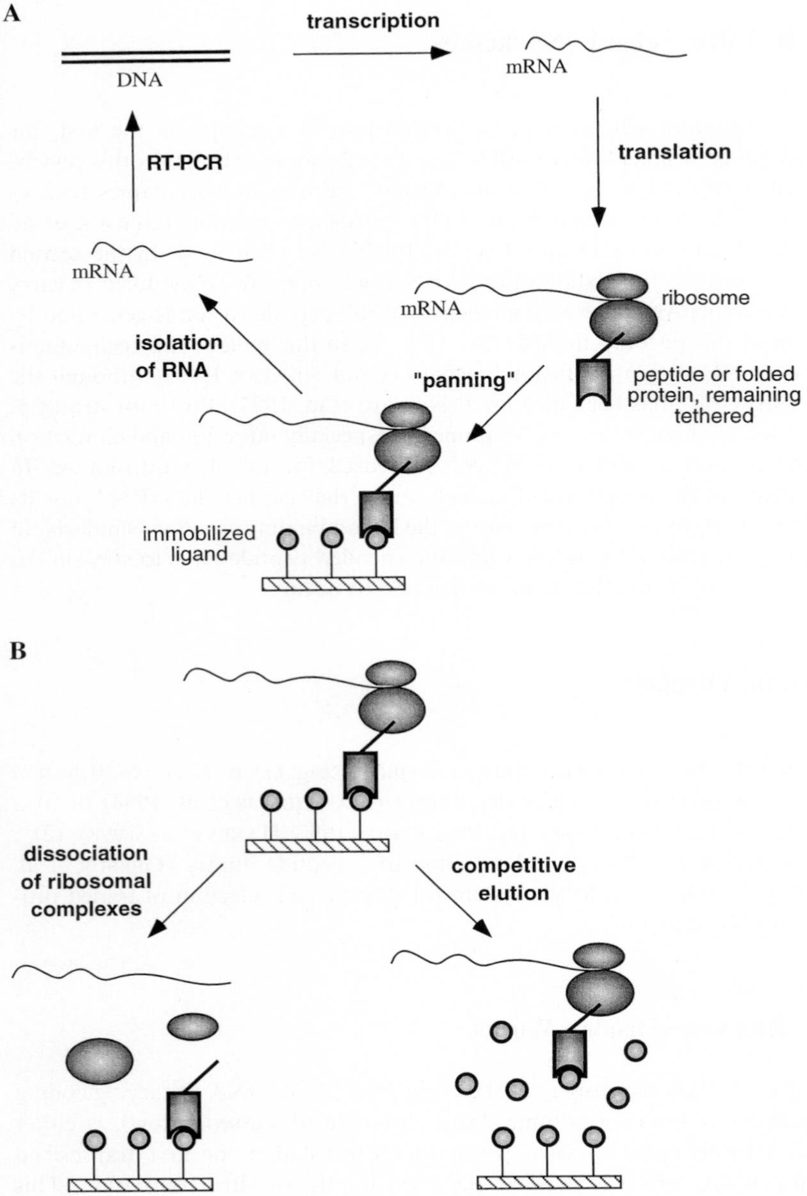

directly, by removing $Mg^{2+}$ with an excess of EDTA and thus causing dissociation of all bound complexes, or first by competitive elution of ribosomal complexes with free ligand followed by RNA isolation only from eluted complexes (Fig. 1B). On the one hand, the latter approach can be advantageous, because the RNA is isolated only from those ribosomal complexes which contain a functional ligand binding protein. But on the other hand, this approach might be difficult to apply for very tight binders.

## 4.2 Construction of a DNA Library

The ribosome display construct (Fig. 3) contains, at the DNA level, a T7 promoter for efficient transcription to mRNA. On the RNA level, the construct contains a prokaryotic ribosome binding site or a Kozak sequence, depending on the translation system used, followed by the protein coding sequence without a stop codon. In a prokarytic cell-free translation system the presence of a stop codon would result in the binding of the release factors (Grentzmann et al. 1995; Tuite and Stansfield 1994; Tate and Brown 1992) and the ribosome recycling factor (Janosi et al. 1994) to the mRNA-ribosome-protein complexes. This in turn would lead to the hydrolysis of peptidyl-tRNA between the 3′-ribose and the last amino acid of the polypeptide by the peptidyltransferase center of the ribosome (Tate and Brown 1992) (Fig. 4A). A similar mechanism is also operative in eukaryotic systems (Frolova et al. 1994; Zhouravleva et al. 1995). No equivalent to the prokaryotic ribosome recycling factor has been identified in eukaryotes so far. Obviously, no stop codon must be present in order to keep mRNA and the encoded protein in the ribosomal complexes. However, there is a backup-system in *E. coli*, involving the 10Sa-RNA (Ray and Apirion 1979), a stable bacterial RNA with tRNA-like structure (Komine et al. 1994). A polypeptide translated in vivo from mRNA without stop codon is modified by COOH-terminal addition of a peptide tag, encoded by the 10Sa-RNA (Tu et al. 1995) and subsequently released from the ribosome (Fig. 4B). The released protein tagged with this sequence is finally degraded by a tail-specific protease (Keiler et al. 1996).

In ribosome display constructs, the open reading frame coding for the protein comprises two portions: the $NH_2$-terminal part, which codes for the polypeptide to

**Fig. 1. A** Principle of ribosome display for screening protein libraries for ligand binding. A DNA library containing all important features necessary for ribosome display (for details see text) is first transcribed to mRNA and after its purification, mRNA is translated in vitro. Translation is stopped by cooling on ice, and the ribosome complexes are stabilized by increasing the magnesium concentration. Ribosomal complexes are affinity selected from the translation mixture by the native, newly synthesized protein binding to immobilized ligand. Nonspecific ribosome complexes are removed by intensive washing, and mRNA is isolated from the bound ribosome complexes, reverse transcribed to cDNA, and cDNA is then amplified by PCR. This DNA is then used for the next cycle of enrichment, and a portion can be analyzed by cloning and sequencing and/or by ELISA or RIA. **B** Two methods for mRNA isolation from bound ribosomal complexes. The bound ribosomal complexes can either be dissociated by an excess of EDTA and then RNA is isolated, or they can first be eluted specifically with free ligand followed by RNA isolation

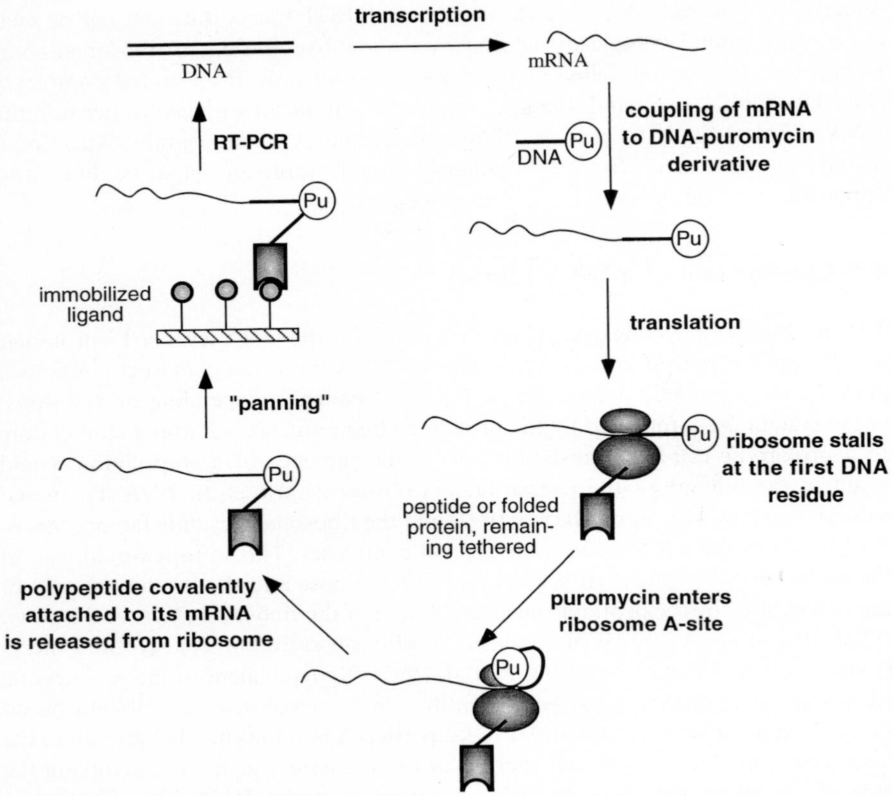


**Fig. 2.** Principle of RNA-peptide-fusion strategy for screening peptide libraries. A DNA peptide library is first transcribed to mRNA, and after its purification mRNA is coupled to a DNA-puromycin derivative. The mRNA-DNA-puromycin derivative is purified and used for in vitro translation. The ribosome stalls at the first DNA residue and puromycin from the translated mRNA enters the ribosomal A-site, where it is covalently linked to the translated peptide with the help of the peptidyl-transferase center. Such an RNA-peptide-fusion no longer needs a ribosome. The desired RNA-peptides are affinity selected from the mixture by binding of the peptide to immobilized ligand. After intensive washing the bound RNA-peptides are isolated, reverse transcribed to cDNA, and cDNA is then amplified by PCR. This DNA is then used for the next cycle of enrichment, and a portion can be analyzed by cloning and sequencing and/or by ELISA or RIA

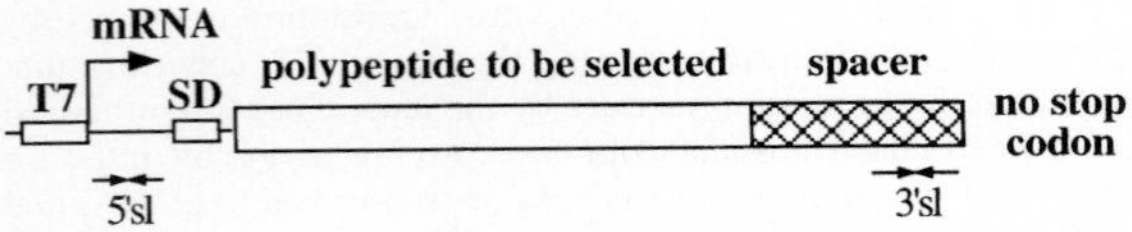


**Fig. 3.** The scFv construct used for prokaryotic ribosome display. *T7* denotes the T7 promoter, *SD* the ribosome binding site, *spacer* the part of the protein construct connecting the folded scFv to the ribosome, *5′*sl and *3′*sl the stem loops on the 5′- and 3′-ends of the mRNA. The *arrow* indicates the transcriptional start

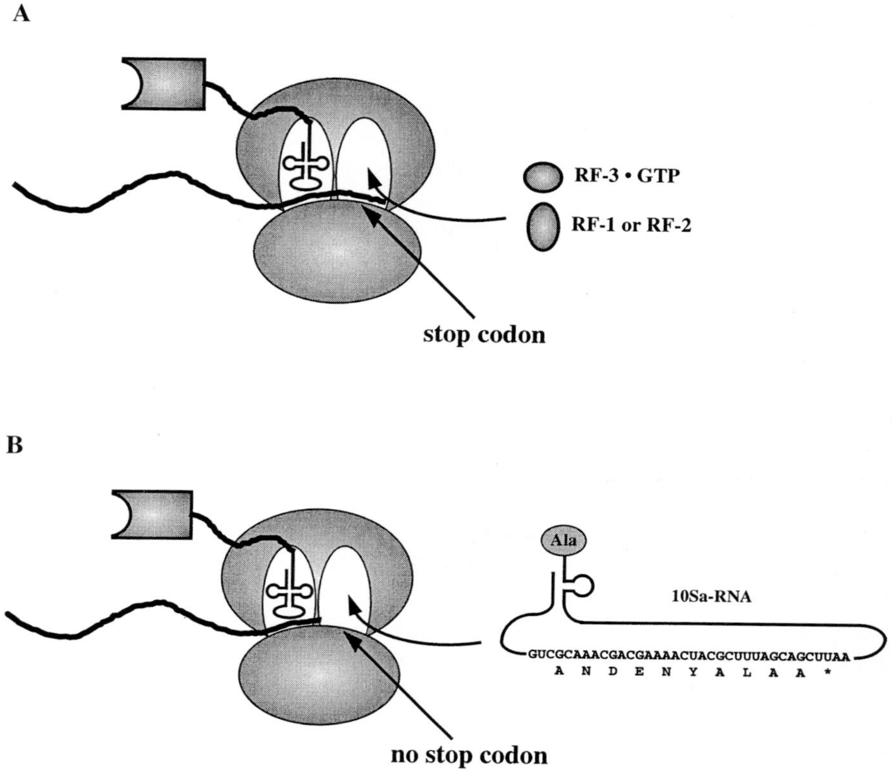


**Fig. 4A,B.** Role of stop codon and 10Sa-RNA in *E. coli*. **A** Role of stop codon. A complex of two release factors (proteins), RF-1 and RF-3 or RF-2 and RF-3, binds in place of tRNA, when a stop codon is encountered. The release factor RF-1 recognizes the stop codons UAA or UAG while factor RF-2 recognizes the stop codons UAA or UGA. The binding of the release factor complex results in hydrolysis of peptidyl tRNA in the ribosome and the protein is released. **B** Role of 10Sa-RNA. Translation of mRNA without a stop codon results in the binding of 10Sa-RNA to the A-site of the ribosome. First alanine, which is the acyl group carried by this RNA, is added to the truncated protein. Then, this RNA is taken to be a messenger RNA, resulting in coupling of a peptide tag encoded by the 10Sa-RNA with the sequence indicated. Because this tag ends with a stop codon, the protein is released normally and then degraded by a protease specific for this COOH-terminal tag

be selected (the library), and the COOH-terminal part, which is constant and serves as a spacer. The spacer has several functions: (1) it tethers the synthesized protein to the ribosomes by maintaining the covalent bond to the tRNA which is bound at the P-site of the ribosome, (2) it keeps the synthesized polypeptide outside the ribosome and allows it to fold and to interact with ligands, despite the fact that the ribosome itself is thought to cover about 20–30 amino acid residues of the emerging polypeptide, and (3) it may slow down protein synthesis, since the spacer can contain rare codons, mRNA secondary structures or other stalling sequences (Mattheakis et al. 1996). However, no beneficial effect of these translation retarding features for display efficiency has been experimentally demonstrated.

A number of different spacers of various lengths have been used. For peptide libraries, spacers of 85 (Mattheakis et al. 1994), 121 (Mattheakis et al. 1996) and 72 amino acid residues (Gersuk et al. 1997) were reported. For protein display, spacers of 88–116 amino acid residues in length were used and found to increase the efficiency of *E. coli* ribosome display with the length of the spacer (Hanes and Plückthun 1997), and in a rabbit reticulocyte system a kappa domain of an antibody served as a spacer of 103 amino acids (He and Taussig 1997).

At the RNA level, additional important features which should be present in the ribosome display construct are 5′- and 3′-stem loops. They are known to stabilize mRNA against RNases and therefore increase the half life of mRNA in vivo as well as in vitro (Belasco and Brawerman 1993). The *E. coli* S-30 extract, which is used for in vitro translation during ribosome display, contains high RNase activity. The introduction of the 5′-stem-loop, derived from the T7 gene 10 upstream region directly at the beginning of the mRNA, and the introduction of the 3′-stem-loop, derived from the terminator of the *E. coli* lipoprotein, into the ribosome display construct were found to improve mRNA stability and therefore increased the efficiency of ribosome display approximately 15-fold (Hanes and Plückthun 1997). A similar improvement was observed when using the analogous 5′-stem-loop and the 3′-stem-loop derived from the early terminator of phage T3 (Hanes and Plückthun 1997).

A ribosome display library template can be conveniently prepared by ligation of a DNA library to the spacer and subsequent amplification of the ligation mixture in two PCRs with two pairs of oligonucleotides, which introduce all above-mentioned features important for ribosome display (e.g. Hanes and Plückthun 1997).

## 4.3 In Vitro Translation

The DNA library can either be directly converted to a ribosome-bound polypeptide library by coupled in vitro transcription-translation (Mattheakis et al. 1994; He and Taussig 1997), or mRNA can be first prepared by in vitro transcription and subsequently used for in vitro translation (Hanes and Plückthun 1997; Gersuk et al. 1997). The coupled system is simpler than the uncoupled one, but it was observed that the efficiency of the coupled *E. coli* system is much lower than the uncoupled system (Hanes et al., unpublished experiments). Another disadvantage of the coupled system is an incompatibility of the redox requirements of transcription and translation when displaying proteins containing disulfide bridges. T7 RNA polymerase, which is necessary for transcription in this system, is usually stabilized by β-mercaptoethanol, which competes with disulfide bond formation. This problem may in principle be overcome by preparing T7 RNA polymerase without reducing agent, but the enzyme's activity must then be carefully monitored.

Translation is usually performed at 37°C when using the *E. coli* in vitro system (Mattheakis et al. 1994; Hanes and Plückthun 1997). Despite the general tendency of proteins to fold with higher efficiency at lower temperature in vitro, the yield of functional molecules from in vitro translation was indeed found to be higher at 37°C. This may be due to the action of chaperones in the extract, and it is a complicated function of the temperature-dependence of translation, folding, RNases and perhaps proteases. For in vitro translation in eukaryotic systems, lower temperatures are usually used, for example 30°C in the rabbit reticulocyte system (He and Taussig 1997) or even 27°C in the wheat germ system (Gersuk et al. 1997).

The time of translation is also an important variable and is more critical for uncoupled systems. At physiological temperatures, the absence of a stop codon is not sufficient to keep the mRNA and its encoded protein complexed to the ribosome forever. An in vitro translation of truncated lysozyme mRNA in a wheat germ system resulted only in free protein, and no protein present in the ribosomal fraction, after 80 min of translation (Haeuptle et al. 1986). The translated protein was only observed to be present in the ribosomal fraction after shortening of the translation time to 60 min, and its concentration increased when the translation was performed for only 30 min (Haeuptle et al. 1986). The translation necessary for ribosome display with an uncoupled wheat germ system was performed for 15 min at 27°C (Gersuk et al. 1997), and the optimal time for the uncoupled *E. coli* system is usually not longer than 10 min at 37°C (Hanes and Plückthun 1997). However, the complexes are very stable, as soon as they are cooled to 4°C (Hanes et al., unpublished experiments). In a coupled transcription-translation system mRNA is continuously produced and therefore the reaction time can be extended to 30–60 min (Mattheakis et al. 1996; He and Taussig 1997). Too long a translation, on the other hand, may lead to the depletion of some crucial component necessary for translation or transcription or the accumulation of low molecular weight compounds inhibiting translation (Jermutus et al. 1998), resulting in a subsequent decrease of the amount of ribosomal complexes.

Several additional components can be used during in vitro translation which can improve the efficiency of ribosome display. RNasin, an inhibitor of RNases, was added during in vitro translation in the wheat germ system (Gersuk et al. 1997), but it was not reported whether it had any effect. RNasin was found to have no influence during in vitro translation in an *E. coli* system (Hanes and Plückthun 1997). However, vanadyl ribonucleoside complexes (Berger and Birkenmeier 1979; Puskas et al. 1982), general RNase inhibitors which should act as transition state analogs, were found to increase the efficiency of *E. coli* ribosome display when used during in vitro translation (Hanes and Plückthun 1997).

For displaying folded disulfide containing proteins such as scFv fragments of antibodies, eukaryotic protein disulfide isomerase (PDI), which catalyzes disulfide bond formation and rearrangement, was found to increase the efficiency of *E. coli* ribosome display three-fold when used during translation (Hanes and Plückthun 1997) (Table 1). The elimination of the 10Sa-RNA (the product of the *ssrA* gene) by an antisense oligonucleotide, which is responsible for tagging (see Sect. 4.2) and releasing truncated peptides from *E. coli* ribosomes (Ray and Apirion 1979; Komine et al. 1994; Keiler et al. 1996), yielded a four-fold improvement of ribosome display when using an antisense DNA oligonucleotide directed against this RNA (Hanes and Plückthun 1997) (Table 1).

**Table 1.** Summary of improvements for increasing the efficiency of ribosome display

| mRNA structure (5′ and 3′-loops) | Additives[a] | Yield of mRNA after one round of affinity selection | | |
|---|---|---|---|---|
| | | Percent of input mRNA | Number of molecules[b] | Relative amount |
| – | No | 0.001 | $1.3 \times 10^{8}$ | 1 |
| + | No | 0.015 | $2.0 \times 10^{10}$ | 15 |
| + | PDI | 0.045 | $5.9 \times 10^{10}$ | 45 |
| + | Anti 10Sa-RNA[c] | 0.060 | $7.9 \times 10^{10}$ | 60 |
| + | PDI, anti 10Sa-RNA | 0.200 | $2.6 \times 10^{11}$ | 200 |

PDI, protein disulfide isomerase.
[a] 0.1% VRC during translation and 50 mM $Mg^{2+}$ during affinity selection were used in all experiments.
[b] Number of molecules isolated from 1 ml reaction.
[c] Antisense oligonucleotide

## 4.4 Affinity Selection of Ribosomal Complexes

After in vitro translation, the reaction is stopped by rapid cooling on ice. In the eukaryotic system, cycloheximide can also be added (Gersuk et al. 1997), but the effect of this compound on the efficiency of the ribosome display has not been reported. The mixture is also diluted with a buffer containing magnesium acetate, which is present during the whole affinity selection. Concentrations of 5 mM magnesium acetate in a wheat germ system (Gersuk et al. 1997) or 10 mM magnesium acetate in an *E. coli* system (Mattheakis et al. 1994) were used. Much higher concentrations (50 mM magnesium acetate) were found to be optimal and improved the efficiency of the *E. coli* ribosome display several-fold (Hanes and Plückthun 1997). A possible explanation for the need for high magnesium concentrations is that $Mg^{2+}$ binds to the phosphates of the ribosomal RNA, the mRNA, and the peptidyl tRNA, thus stabilizing ribosome complexes. In the absence of magnesium, ribosome complexes may dissociate. No magnesium was used during the reported affinity selection in a rabbit reticulocyte system (He and Taussig 1997). It would be interesting to see if any improvement resulted from addition of $Mg^{2+}$ to this system.

Chloramphenicol, an antibiotic which inhibits bacterial protein synthesis by binding to the 23S ribosomal RNA in the peptidyl transferase center, has been used throughout the entire affinity selection processes in an *E. coli* system (Mattheakis et al. 1994). However, in a direct comparison, chloramphenicol was found to have no influence on the efficiency of *E. coli* ribosome display (Hanes and Plückthun 1997).

While Mattheakis et al. (1994) preparatively separated the ribosomal complexes by centrifugation through a sucrose cushion prior to affinity selection, all other reports used the translation mixture directly for panning (Hanes and Plückthun 1997; He and Taussig 1997; Gersuk et al. 1997). In a direct comparison, no improvement by the isolation of ribosomal complexes through a sucrose cushion was found (Hanes and Plückthun 1997).

Affinity selection can be performed by using either ligands immobilized on a surface (such as panning tubes or microtiter wells) or biotinylated ligands bound to the ribosome-bound proteins which are subsequently captured by streptaridin-coated magnetic beads. After extensive washing with a magnesium-containing buffer, mRNA can be isolated either from ribosome complexes dissociated with EDTA, or from complexes specifically eluted with an excess of a free ligand (Fig. 1B). The isolated mRNA is then used for RT-PCR, and the DNA thus obtained can be used for the next cycle of ribosome display. A portion of the DNA can be analyzed by cloning and sequencing and/or by ELISA or RIA after each round of ribosome display. When magnetic beads are used for selection, RT-PCR can also be directly performed with a portion of the beads (HE and TAUSSIG 1997).

The efficiency of ribosome display can also be improved by decreasing the nonspecific binding. Supplementing the diluted translation mixture before affinity selection with 2% sterilized milk and/or 0.2% heparin eliminates much of the nonspecific binding, perhaps by preventing binding of ribosome complexes to the panning tube surface or to magnetic beads, and heparin probably acts also as an RNase inhibitor (HANES et al. 1998).

## 5 Applications of Ribosome Display

### 5.1 Display of Peptides on Ribosomes

An *E. coli* ribosome display system was used for displaying peptides from a decamer random library. This library was selected for binding to the monoclonal antibody D32.39, which has been raised to bind dynorphin B, a 13-residue opioid peptide, with 0.29 nM affinity (MATTHEAKIS et al. 1994). A library of $10^{12}$ DNA molecules was used for ribosome display using coupled in vitro transcription-translation. After five rounds of ribosome display, several different peptides ranging from 7.2 to 140 nM affinity to the antibody were found. However, a peptide similar to dynorphin B, or any peptides possessing a similar affinity, were not obtained.

A wheat germ ribosome display system was used for displaying a 20-mer random library, which was selected for binding to a prostate-specific antigen (PSA) (GERSUK et al. 1997). After four rounds of selection, several peptides showing higher affinity to PSA than to bovine serum albumin or gelatin were isolated, but no quantitative data were reported.

### 5.2 Display of Folded Proteins on Ribosomes

Two scFv fragments of an antibody were used as a model system: scFv-hag, derived from the antibody 17/9 (specific for hemagglutinin peptide) (SCHULZE-GAHMEN et al. 1993), and scFv-AL2 (specific for ampicillin) (KREBBER et al. 1996). mRNAs of these two scFvs were mixed in a ratio of $1{:}10^8$ (scFv-hag:scFv-amp) and applied for affinity selection by using ribosome display on a hag-surface (HANES and PLÜCKTHUN 1997). After five rounds of selection, a pool of enriched sequences was cloned and single clones were analyzed. Of 20 scFvs, 18 were scFv-hag and two were scFv-amp, demonstrating that the $10^9$-fold enrichment was successful. The average enrichment per cycle of ribosome display was thus about 100-fold, and this number is now known to depend both on the antibody, antigen and type of surface (Hanes et al., unpublished experiments). All 18 scFvs with the anti-hag sequence were analyzed by RIA, and it was shown that all but one of them bind hag peptide and can be inhibited by it. Sequence analysis showed that all of them had mutated during the five cycles of ribosome display and possessed between zero and four amino acid substitutions with respect to the original scFv-hag. All changes were independent of each other. It was also shown that from a binary mixture of scFv-hag and scFv-amp mRNAs, mixed in a ratio of 1:1, either scFv could be enriched, depending on which antigen was used for affinity selection (HANES and PLÜCKTHUN 1997).

*E. coli* ribosome display was applied to affinity selection of scFv antibody fragments from a diverse library generated from mice immunized with a variant peptide of the transcription factor GCN4 dimerization domain (HANES et al. 1998). The *E. coli* ribosome display system using uncoupled in vitro translation and all the improvements reported by HANES and PLÜCKTHUN (1997) were used. After three rounds of ribosome display, an enriched pool of scFv genes were cloned and single clones were analyzed by RIA. Twenty-six different scFvs binding to a GCN4-variant peptide were isolated. Several different scFvs were selected, but the largest group of 22 scFvs was closely related to each other and differed in zero to five amino acid residues with respect to their consensus sequence, the likely common progenitor. The other four scFvs were different from each other and also from the group of closely related 22 scFvs, and showed lower affinity to the GCN4-variant peptide than the 22 related scFvs, based on RIA analysis. The best scFv was found among the related ones and had a dissociation constant of $(4 \pm 1) \times 10^{-11}$M, measured in solution. The scFv identical to the consensus sequence, a likely common progenitor of the 22 related scFvs, was identified and had a dissociation constant of only $(2.6 \pm 0.1) \times 10^{-9}$M. Detailed analysis showed that for the 65-fold higher affinity of the best scFv to the antigen only one amino acid change in the "progenitor" scFv was responsible. It was also shown that this high-affinity scFv was selected from mutations occurring in vitro during ribosome display rounds and that it was not present in the original library. Thus, this selected scFv had evolved throughout the rounds of ribosome display (HANES and PLÜCKTHUN 1997). The in vitro selected scFvs could be functionally expressed in the *E. coli* periplasm with good yields, or prepared by in vitro refolding in equally good yields.

Rabbit reticulocyte ribosome display was also used to select a scFv derivative of an antibody, of the type $V_H$-linker-$V_L$-$C_L$, binding to progesterone. A mini-library was prepared by mixing the DNA coding for this construct derived from the progesterone specific antibody $DB3^R$ (carrying the mutation Trp100Arg in $V_H$ which does not influence antigen binding), and a DNA of several mutant scFvs in position H35 which do not bind this antigen (HE and TAUSSIG 1997). The enrichment was analyzed by DNA sequencing of the pool. No antigen binding data (RIA or ELISA) of the translated protein have been reported to date. $DB3^R$ DNA diluted $10^2$- to $10^6$-fold with mutant $DB3^{H35}$ DNA was applied for ribosome display using progesterone-BSA, covalently immobilized on magnetic beads. After one round of ribosome display $DB3^R$ was reported to be the dominant species, recovered from $10^2$- to $10^4$-fold dilution, and comprising about 50% when enriched from a $10^5$-fold diluted mixture. The reported nominal enrichment for one cycle of ribosome display was therefore about $10^4$–$10^5$, even though the translation mixture contained 2 mM DTT, and the ribosomes were not stabilized by $Mg^{2+}$. Yet, in a direct comparison of the eukaryotic and the *E. coli* ribosome display system with the same scFv fragment, lower enrichments were found for the eukaryotic system (Hanes et al., submitted).

## 6 RNA-Peptide Fusion

The selection principle of the RNA-peptide fusion approach is directly related to the ribosome display technology, but uses a puromycin-tagged mRNA and several additional steps to achieve covalent coupling of mRNA to its encoding polypeptide (Fig. 2). The method has so far been applied using the rabbit reticulocyte in vitro translation system (Roberts and Szostak 1997; Nemoto et al. 1997). The RNA-peptide fusion construct consists of the mRNA coding for the peptide sequence, fused either to a DNA spacer of the sequence $dA_{27}dCdC$ coupled to puromycin (Roberts and Szostak 1997) or to a DNA–RNA hybrid spacer of 125 deoxynucletides and four ribonucleotides coupled to puromycin (Nemoto et al. 1997). The principle of this system is similar to ribosome display: an mRNA–DNA–puromycin hybrid is used for in vitro translation in a rabbit reticulocyte system. When the ribosome reaches the DNA portion of the template, translation stalls. At this point the ribosome complex can either dissociate, or puromycin, which is part of the template, can enter the ribosome and attach itself to the synthesized peptide. In this way the genotype, mRNA, can be directly attached to the phenotype, its encoded peptide.

For testing the system, the template encoding a myc-epitope was used. It was shown that the RNA-myc-peptide fusion can be isolated from an in vitro translation mixture by immunoprecipitation with a monoclonal antibody recognizing a myc-epitope (Roberts and Szostak 1997). The myc-peptide-mRNA fusion was also enriched by immunoprecipitation from a mixture prepared by translation of myc-encoding template, diluted with template encoding a random peptide pool. The reported enrichment factor was 20- to 40-fold. Nonspecific RNA-peptide fusions (peptides coupled not to their encoding RNA) were not observed, and thus the enrichment is probably limited by nonspecific binding of peptides and unprotected RNA or DNA to the target. About 1% of RNA was converted to the RNA-peptide fusion (Roberts and Szostak 1997).

## 7 Conclusions and Perspectives

Two closely related in vitro selection methods for screening of peptide and protein libraries, ribosome display and RNA-peptide fusion, have been reported so far. Both are based on a cell-free translation system, and all steps are performed in vitro without using cells in any step. The phenotype, the synthesized protein or peptide, is attached to its genotype, encoded by mRNA, either by complexing it with the ribosome in the ribosome display system or, in the RNA-peptide fusion method, by subsequent covalent attachment of the synthesized peptide to its puromycin-derivatized mRNA. The ribosome display system was shown be effective in the selection of peptides and also of functional, folded proteins from complex libraries, while selection experiments with the RNA-peptide fusion system have so far been reported for peptides.

The main advantage of in vitro selection methods is, as already mentioned, that no cloning is necessary and therefore very large libraries can be used for screening and selection. The ribosome display system can be performed very rapidly; one cycle of selection can be achieved within one day, which compares well with in vivo selection methods, when a library has to be prepared. In the RNA-peptide fusion approach the mRNA-puromycin derivative must be synthesized after each cycle, and therefore this method is somewhat slower than ribosome display. Another advantage of in vitro selection methods is the automatic introduction of mutations during the procedure, when non-proofreading polymerases are used, and thus proteins can also affinity-mature during selection. It appears that in vitro selection methods have a great utility in compound identification and optimization, and they can thus help to answer fundamental questions of protein structure and evolution.

## References

Belasco JG, Brawerman G (1993) Control of messenger RNA stability. Academic press Inc., San Diego

Berger SL, Birkenmeier CS (1979) Inhibition of intractable nucleases with ribonucleoside–vanadyl complexes: isolation of messenger ribonucleic acid from resting lymphocytes. Biochemistry 18:5143–5149

Crameri A, Whitehorn EA, Tate E, Stemmer WP (1996) Improved green fluorescent protein by molecular evolution using DNA shuffling. Nature Biotech. 14:315–319

Dower WJ, Cwirla SE (1992) In: Chang DC, Chassy BM, Saunders JA and Sowers AE (eds) Guide to electroporation and electrofusion. Academic Press, San Diego, pp 291–301

Fields S, Sternglanz R (1994) The two-hybrid system: an assay for protein–protein interactions. Trends in Genetics 10:286–292

Frolova L, Le Goff X, Rasmussen HH, Cheperegin S, Drugeon G, Kress M, Arman I, Haenni AL, Celis JE, Philippe M, Justesen J, Kisselev LL (1994) A highly conserved eukaryotic protein family possessing properties of polypeptide chain release factor. Nature 372:701–703

Gersuk GM, Corey MJ, Corey E, Stray JE, Kawasaki GH, Vessella RL (1997) High-affinity peptide ligands to prostate-specific antigen identified by polysome selection. Biochem Biophys Res Commun 232:578–582

Grentzmann G, Brechemier-Baey D, Heurgue-Hamard V, Buckingham RH (1995) Function of polypeptide chain release factor RF-3 in *Escherichia coli.* RF-3 action in termination is predominantly at UGA-containing stop signals. J Biol Chem 270:10595–10600

Haeuptle MT, Frank R, Dobberstein B (1986) Translation arrest by oligodeoxynucleotides complementary to mRNA coding sequences yields polypeptides of predetermined length. Nucleic Acids Res 14:1427–1448

Hanes J, Plückthun A (1997) In vitro selection and evolution of functional proteins using ribosome display. Proc Natl Acad Sci USA 91:4937–4942

Hanes J, Jermutus L, Weber-Bornhauser S, Bosshard HR, Plückthun A (1998) Ribosome display efficiently selects and evolves high-affinity antibodies in vitro from immune libraries. Proc Natl Acad Sci USA 95:14130–14135

He M, Taussig MJ (1997) Antibody-ribosome-mRNA (ARM) complexes as efficient selection particles for in vitro display and evolution of antibody combining sites. Nucleic Acids Res 25:5132–5134

Janosi L, Shimizu I, Kaji A (1994) Ribosome recycling factor (ribosome releasing factor) is essential for bacterial growth. Proc Natl Acad Sci USA 91:4249–4253

Jermutus L, Ryabova L, Plückthun A (1998) Recent advances in producing and selecting functional proteins by cell-free translation. Curr Opin Biotechnol 9:534–548

Kawasaki G (1991) Screening randomized peptides and proteins with polysomes. PCT Int Appl WO 91/05058

Keiler KC, Waller PR, Sauer RT (1996) Role of a peptide tagging system in degradation of proteins synthesized from damaged messenger RNA. Science 271:990–993

Komine Y, Kitabatake M, Yokogawa T, Nishikawa K, Inokuchi H (1994) A tRNA-like structure is present in 10Sa RNA, a small stable RNA from Escherichia coli. Proc Natl Acad Sci USA 91:9223–9227

Kraus JP, Rosenberg LE (1982) Purification of low-abundance messenger RNAs from rat liver by immunoadsorption. Proc Natl Acad Sci USA 79:4015–4019

Krebber A, Bornhauser S, Burmester J, Honegger A, Willuda J, Bosshard HR, Plückthun A (1996) Reliable cloning of functional antibody variable domains from hybridomas and spleen cell repertoires employing a reengineered phage display system. J Immunol Meth 201:35–55

Low NM, Holliger P, Winter G (1996) Mimicking somatic hypermutation: affinity maturation of displayed on bacteriophage using a bacterial mutator strain. J Mol Biol 260:359–368

Mattheakis LC, Bhatt RR, Dower WJ (1994) An in vitro polysome display system for identifying ligands from very peptide libraries. Proc Natl Acad Sci USA 91:9022–9026

Mattheakis LC, Dias JM, Dower WJ (1996) Cell-free synthesis of peptide libraries displayed on polysomes. Methods in Enzymology 267:195–207

Moore JC, Arnold FH (1996) Directed evolution of a para-nitrobenzyl esterase for aqueous-organic solvents. Nature Biotech 14:458–467

Nemoto N, Miyamoto-Sato E, Husimi Y, Yanagawa H (1997) In vitro virus: bonding of mRNA bearing puromycin at the 3′-terminal end to the C-terminal end of its encoded protein on the ribosome in vitro. FEBS Lett 414:405–408

Payvar F, Schimke RT (1979) Improvements in immunoprecipitation of specific messenger RNA. Isolation of highly purified conalbumin mRNA in high yield. Eur J Biochem 101:271–282

Phizicky EM, Fields S (1995) Protein–protein interactions: methods for detection and analysis. Microbiol Rev 59:94–123

Puskas RS, Manley NR, Wallace DM, Berger SL (1982) Effect of ribonucleoside-vanadyl complexes on enzyme-catalyzed reactions central to recombinant deoxyribonucleic acid technology. Biochemistry 21:4602–4608

Rader C, Barbas CF 3rd. (1997) Phage display of combinatorial antibody libraries. Curr Opin Biotechnol 8:503–508

Ray BK, Apirion D (1979) Characterization of 10 S RNA: a new stable RNA molecule from Escherichia coli. Mol Gen Genet 174:25–32

Roberts RW, Szostak JW (1997) RNA-peptide fusions for the in vitro selection of peptides and proteins. Proc Natl Acad Sci USA 94:12297–12302

Schechter I (1973) Biologically and chemically pure mRNA coding for a mouse immunoglobulin L-chain prepared with the aid of antibodies and immobilized oligothymidine. Proc Natl Acad Sci USA 70:2256–2260

Schulze-Gahmen U, Rini JM, Wilson IA (1993) Detailed analysis of the free and bound conformations of an antibody. X-ray structures of Fab 17/9 and three different Fab-peptide complexes. J Mol Biol 234:1098–1118

Tate WP, Brown CM (1992) Translational termination: "stop" for protein synthesis or "pause" for regulation of gene expression. Biochemistry 31:2443–2450

Tu GF, Reid GE, Zhang JG, Moritz RL, Simpson RJ (1995) C-terminal extension of truncated recombinant proteins in Escherichia coli with a 10Sa RNA decapeptide. J Biol Chem 270:9322–9326

Tuite MF, Stansfield I (1994) Termination of protein synthesis. Mol Biol Reports 19:171–181

Vaughan TJ, Williams AJ, Pritchard K, Osbourn JK, Pope AR, Earnshow JC, McCafferty J, Hodits RA, Wilton J, Johnson KS (1996) Human antibodies with sub-nanomolar affinities isolated from a large non-immunized phage display library. Nature Biotechnology 14:309–314

Yang WP, Green K, Pinz-Sweeney S, Briones AT, Burton DR, Barbas 3rd. CF (1995) CDR walking mutagenesis for the affinity maturation of a potent human anti-HIV-1 antibody into the picomolar range. J Mol Biol 254:392–403

Zhouravleva G, Frolova L, Le Goff X, Le Guellec R, Inge-Vechtomov S, Kisselev L, Philippe M (1995) Termination of translation in eukaryotes is governed by two interacting polypeptide chain release factors, eRF1 and eRF3. EMBO J 14:4065–4072

# Aptamers as Tools in Molecular Biology and Immunology

M. Famulok[1] and G. Mayer[2]



## 1 Introduction

In 1990 Tuerk and Gold introduced the first RNA aptamer for bacteriophage T4 DNA polymerase, obtained by a new combinatorial technique which they designated as SELEX (systematic evolution of ligands by exponential enrichment). In parallel, Ellington and Szostak (1990) showed that it is also possible to select RNA aptamers which are able to specifically complex organic molecules of low molecular weight, thus serving as receptor molecules based on nucleic acids rather than proteins. Since then, considerable progress has been achieved in the field of in vitro selection of combinatorial nucleic acid libraries, which demonstrates its impressive potential as a tool in molecular biology, diagnostics, molecular medicine, drug discovery, and bio-organic chemistry. Today, the SELEX process has been applied to more than a hundred different target molecules, and aptamers are known for almost every kind of targets such as organic dyes, amino acids, biological cofactors, antibiotics, peptides and proteins or even whole viruses (Bell et al. 1998; Gal et al. 1998; Ellington and Osborne 1997; Kraus et al. 1998; Yang et al. 1998; Eaton 1997; Pan et al. 1995), showing that aptamers can be obtained for almost any desired target whether complex or small.

[1] Kekule-Institut für Organische Chemie and Biochemie, Gerhard-Domagk-Str. 1, D-53121 Bonn, Germany
[2] Institut für Biochemie der LMU München, Feodor-Lynen-Str. 25, D-81377 Munich, Germany

The isolation of specific antagonists for proteins which are involved in disease processes is one of the major goals in pharmacological research. Drug discovery has been greatly facilitated by computer-assisted drug design and various screening strategies of diverse combinatorial libraries of small molecules, peptides, Fab fragments, and antibodies. The SELEX technology provides a powerful method for the screening of large libraries of oligonucleotides, with diversities of up to $10^{15}$ different molecules, for specific ligand-binding nucleic acids which in many cases have been shown to not only bind a certain target protein, but also to inhibit its biological function. Many isolated aptamers are aimed at possible therapeutic and/or diagnostic applications. Insufficient stability, often cited as the major potential drawback of nucleic acids as therapeutic agents, can easily be overcome by using libraries of chemically modified nucleic acids, such as 2′-fluoro- or 2′-amino-2′-deoxypyrimidine containing nucleic acids. Modifications of that kind have been shown to be compatible with the enzymatic steps of the SELEX process. Other strategies which circumvent the stability problem of RNA or DNA include the so-called mirror-image, or Spiegelmer, approach by exploiting nuclease resistance of the enantiomer of naturally occurring nucleic acids (Klußmann et al. 1996; Nulte et al. 1996).

Various recent examples illustrate the potential of aptamers in affecting cellular processes. In this review we will give an overview on recent progress in oligonucleotide selections and applications of aptamers as potential tools in drug discovery, diagnostics, molecular medicine, and for the dissection of cellular processes of immunological relevance (Table 1).

**Table 1.** Summary of the targeted molecules/viruses used for SELEX experiments

| Target molecule | Dissociation constant (nM) | Possible therapeutic and/or diagnostic application | Reference |
|---|---|---|---|
| IgE | 10.0 | Allergic disease | Wiegand et al. (1996) |
| Anti-acetylcholine autoantibodies | 60.0 | Myasthenia gravis | Lee and Sullenger (1997) |
| IFN-γ | 6.8 | Inflammation and immune response | Kubik et al. (1997) |
| L-selectin | 3.0 | Inflammation | O'Connell et al. (1996) |
| CD4 | N/A | Immune response | Kraus et al. (1998) |
| bFGF | 0.350 | Angiogenesis | Jellinek et al. (1993) |
| KGF | 0.0003 | Epithelial hyperproliferative disease | Pagratis et al. (1997) |
| PDGF | 0.1 | Tumor development | Green et al. (1996) |
| VEGF | 0.14 | Neovascularization | Green et al. (1995) |
| RSV | 40.0 | Viral infection | Pan et al. (1995) |
| HIV-1 RT | 1.0 | Viral replication | Tuerk et al. (1992) Schneider et al. (1995) |
| HIV-1 integrase | 10.0 | Viral replication | Allen et al. (1995) |
| HIV-1 rev | N/A | Viral replication | Giver et al. (1993) |
| α-thrombin | 25.0 | Thrombosis | Bock et al. (1992), Kubik et al. (1994), Latham et al. (1994) |
| Activated protein C | 110.0 | Thrombosis | Gal et al. (1998) |
| hNE | N/A | Inflammation | Charlton et al. (1997) |
| PTPase | 18.0 | Oncogenesis, viral and cellular regulation | Bell et al. (1998) |
| $PLA_2$ | 118.0 | ARDS, Septic shock | Bridonneau et al. (1998) |

## 2 Aptamers for Antibodies

IgE antibodies play a key role in allergic responses. Aptamer antagonists, single-stranded DNA and 2′-amino-modified RNA molecules which bind human IgE have been isolated by in vitro selection (WIEGAND et al. 1996). Truncated aptamers (designated as IGEL1.2 -2′-amino RNA- and D17.4 -ssDNA-) were shown to competitively inhibit the interaction of human IgE with its FcεRI receptor in rat basophilic leukemia cells transfected with the human IgE receptor. The oligonucleotides block the IgE-mediated release of biogenic amines, such as serotonin, which are important mediators in allergic responses. Because of these properties the selected aptamers may have the potential to serve as a class of new therapeutics for the therapy of allergic diseases.

An aptamer directed against the antibody which plays a decisive role during the development of the autoimmune disease myasthenia gravis was shown to serve as a decoy RNA molecule (LEE and SULLENGER 1997). Myasthenia gravis is a muscular disease which results in progressive muscle weakness. This effect is based on the binding of anti-receptor autoantibodies to acetylcholine receptors in the motor end plate of neuromuscular junctions, affecting failure of muscle response to neuronal impulses (LINDSTROM et al. 1988). The selected aptamer, modified by substitution of the 2′-OH for a 2′-$NH_2$-group at the riboses of pyrimidine residues to improve its resistance against degradation by nucleases, recognizes anti-acetylcholine receptor autoantibodies present in the serum of patients that suffer from myasthenia gravis. Additionally they inhibit the binding of acetylcholine receptors to the autoantibodies. In this way, the undesired antibody-mediated immune response could be inhibited.

DOUDNA et al. (1995) isolated RNA sequences that mimicked a major autoantigenic epitope of the human insulin receptor. The goal of this approach was to obtain an aptamer which could be used in the treatment of patients with extreme insulin resistance type B. The selection was performed with an anti-human insulin receptor mouse monoclonal antibody (designated as MA20) which recognizes the same epitope of the insulin receptor-like autoantibodies present in the sera of these patients. The selected aptamers showed cross-reactivity with the autoantibodies and were shown to serve as RNA decoys, with the effect that they protected the insulin receptor from antibody binding.

## 3 The Interruption of Lymphocyte Signal Transduction Pathways by Aptamers

Aptamers targeted against proteins which are involved in inflammatory processes are of considerable medical interest. Interferon γ (IFN-γ) is an immunoregulatory cytokine which is synthesized and secreted by Th1 and NK cells (TRINCHIERI and

Perussia 1985). It is thought to participate in nearly all phases of immune and inflammatory responses (Farrar and Schreiber 1993). IFN-γ plays crucial roles in promoting inflammatory responses by stimulating macrophage cytotoxicity, enhancing adhesion of vascular endothelial cells, and facilitating lymphocyte extravasation. Kubik et al. (1997) isolated RNA aptamers which bind to IFN-γ with high affinity and specificity. Furthermore, the aptamer inhibits IFN-γ-induced expression of MHC class I and ICAM-1, important proteins in the genesis of inflammatory disease, by human myeloid leukemia cells. The regulation of IFN-γ activity by specific ligands which inhibit the binding of IFN-γ to its receptors may have an important potential for therapeutic applications. This aptamer might be of interest for diagnostic purposes as well.

Lee et al. (1996) reported that several synthetic oligonucleotides can block IFN-γ mediated effects in human cell cultures. Originally these oligonucleotides were designed as antisense-molecules to form a triple helix with the X-$X_2$ box region of the HLA-DRα gene promoter. Because the oligonucleotide is able to inhibit IFN-γ induced up-regulation of MHC class I and II and ICAM-1 expression, Lee et al. (1996) suggested a mechanism similar to that of aptamers for the inhibitory function of the synthetic oligonucleotide, by dissecting the signal transduction pathway through IFN-γ binding. This is an impressive example for a possible application of an oligonucleotide as an inhibitory drug in medicinal chemistry, in this case for inhibition of undesired immune responses after transplantation.

Many cell-cell interactions in the vascular system are regulated by selectins, a family of cell adhesion molecules subdivided into L-selectin, P-selectin, and E-selectin. Selectins are involved in early steps of tissue injury following hypoxemia, reperfusion, or inflammation (Bevilacqua and Nelson 1993; Bevilacqua et al. 1994). Various ligands are known, the most prominent one being silalyl Lewis$^x$, that block the interaction of L-selectin with other cell surface receptors, such as glycosylation-dependent cell adhesion molecule-1 (GlyCAM-1), but none of these ligands is able to discriminate between the three selectin isoforms. The SELEX technology was used to isolate oligonucleotide ligands for L-selectin (O'connell et al. 1996). Nuclease stabilization was achieved by using a 2′-amino-2′-deoxypyrimidine containing oligonucleotide library. The isolated aptamer was shown to bind L-selectin with high affinity and specificity in a calcium-dependent manner and to selectively discriminate between the three selectin isoforms. Furthermore, the aptamer was able to recognize the native protein on cell surfaces.

Another example of aptamers directed against an immunologically relevant target are RNA molecules selected to bind to the CD4 protein. The CD4 antigen interacts with MHC class II on antigen presenting cells (APCs) and plays an essential part in HIV-1 infection (Dagleish et al. 1984). In previous studies it was shown that antibodies directed against CD4 can block both MHC II and HIV-1 complex formation. Using a library of 2′-fluoro modified RNAs, randomized at 36 positions Kraus et al. (1998), isolated aptamers which bind CD4. These aptamers were able to inhibit mixed lymphocyte reactions nearly as well as Fab fragments or IgG antibodies, which may lead to a possible application targeted to manipulations of the immune system.

# 4 Growth Factors as Target Proteins for In Vitro Selection

Growth factors play an important role in cell proliferation, cell migration, tissue remodeling, and wound healing (FOLKMAN and KLAGSBRUN 1987; GOSPODAROWITZ 1991). Angiogenesis, the development of new blood vessels, is often associated with pathological processes such as cancer development and metastasis. For that reason, there is great interest in obtaining antagonists that inhibit growth factor functions. In the following, we give an overview of nuclease-resistant aptamers that have been isolated by in vitro selection for binding to a variety of growth factors, such as basic fibroblast growth factor (bFGF), human keratinocyte growth factor (hKGF), platelet derived growth factor (PDGF), and vascular endothelial growth factor (VEGF).

JELLINEK et al. (1995) reported the isolation of 2′-amino-modified RNA inhibitors of bFGF. This growth factor initiates angiogenesis by binding to either heparin proteoglycans (low-affinity site) or tyrosine kinase receptors (high-affinity site) on the surfaces of endothelial cells (MOSCATELLI 1987). Overexpression of bFGF is correlated with many malignant disorders. The selected aptamer was extensively characterized, and it could be shown that a minimal aptamer, designated as M21A, is sufficient for binding to bFGF, with a dissociation constant of 350pM. M21A inhibits the binding of bFGF to low-affinity sites on CHO cells with an $ED_{50}$ of 1nM and the binding to high-affinity sites with an $ED_{50}$ of 3nM. The bFGF-dependent migration of bovine aortic endothelial cells (BAEs) is inhibited by the aptamer M21A in a dose-dependent manner.

PAGRATIS et al. (1997) selected 2′-amino- and 2′-fluoro-2′-deoxyribonucleotide-modified RNA inhibitors for hKGF. KGF, a member of the FGF family (FINCH et al. 1989), is a basic heparin binding growth factor for epithelial cells (RUBIN et al. 1989). By overexpression of hKGF it was shown that this growth factor participates in diseases like psoriasis and dermatoses with psoriasiform hyperplasia (STAIANO-COICO et al. 1993), inflammatory bowel disease (BRAUCHLE et al. 1996; FINCH et al. 1996), and neoplasia (ISHIL et al. 1994). The isolated hKGF aptamer antagonists bind hKGF with a dissociation constant of up to 0.3pM, which is the tightest binding aptamer reported so far. Compared with the 2′-amino-modified oligonucleotide ligands, the obtained 2′-fluoro-modified RNA ligands showed higher affinities and bioactivities. The aptamers can competitively inhibit hKGF binding to its receptor and inhibits mitogenic activity with $k_i$ values of 92pM. Thus, these nuclease-resistant molecules may be useful for the development of novel pharmaceutical lead compounds for epithelial hyperproliferative disease.

Tumor cell lines produce and secrete PDGF. Therefore, autocrine as well as paracrine effects on tumor stroma and tumor growth can be imagined. Besides the impact of PDGF on tumor development, this growth factor also participates in other proliferative disorders such as glomerulonephritis (IIDA et al. 1991) and arteriosclerosis (LINDNER et al. 1995; LINDNER and REIDY 1995). SELEX-derived ssDNA selected for specific complexing with PDGF showed an inhibition of PDGF function (GREEN et al. 1996). The minimal PDGF binding motifs of the selected aptamers inhibit PDGF interaction with its a- and b-receptors and even specifically inhibit mitogenic effects of PDGF. All isolated aptamers interfered with the PDGF-BB and PDGF-AB isoforms but not with the PDGF-AA isoform, indicating that the interaction is site-specific for the PDGF B-chain.

The aptamer for vascular permeability factor/vascular endothelial growth factor (VPF/VEGF) provides an excellent and very intriguing example for post-SELEX modifications of 2′-aminopyrimidine-modified RNA aptamers (GREEN et al. 1995) (Fig. 1). The isolated high affinity RNA aptamers, which contained the 2′-amino-2′-deoxy modifications only at the pyrimidine nucleotides, could be reduced to a 24 nucleotide minimal motif that is sufficient for VPF/VEGF binding. The minimal aptamer contains 14 ribopurines, and the substitution of ten of them by 2′-O-methyl purine derivatives improved the binding affinity 17-fold and increased the half life up to $t_{1/2} = 131$h. The 2′-ribose-positions that would tolerate a substitution by the 2′-methoxy group were defined by synthesizing the minimal oligonucleotide from 2:1 mixtures of phosphoramidite derivatives of natural purines (RNA-purines) and unnatural 2′-methoxy-2′-deoxy purines. Re-selection of the resulting library of variants of the originally selected aptamer allowed the separation of aptamer variants that showed decreased affinity to VPF/VEGF, as a result of the substitution of some 2′-OH-purine residues for 2′-$OCH_3$-purines from those in which this substitution improved the aptamer binding affinity. These positions were then identified by digestion of radiolabeled modified aptamers at high pH. Under these conditions, selective hydrolysis of the RNA is observed only at the remaining 2′-OH, but not at the 2′-methoxy residues.

The minimal modified aptamer is currently in clinical trials to prove its ability to inhibit angiogenesis and neovascularization, which contributes to the pathology of many angiogenesis-associated diseases.

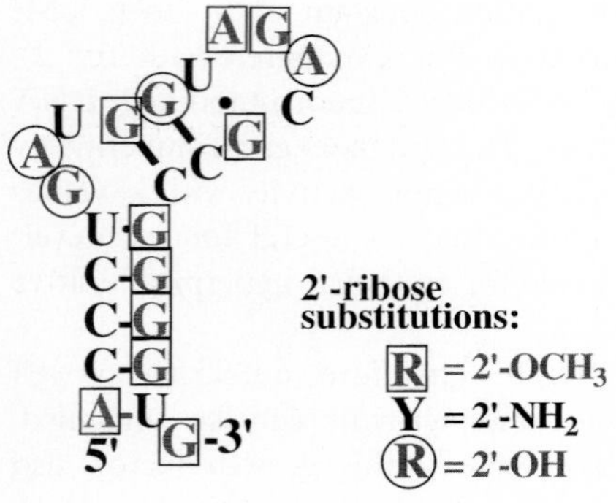


**Fig. 1.** The minimal VEGF aptamer. The post-SELEX modifications of purine bases are highlighted in green. (From GREEN et al. 1995)

## 5 Anti-viral Aptamers

Aptamers deliver potential tools for dissecting the life cycle of viruses by interacting with proteins essential for viral replication. For human immunodeficiency virus-1 (HIV-1) and Rous sarcoma virus (RSV), oligonucleotides have been isolated that can efficiently interrupt essential steps in the viral life cycle, opening up the potential of treating viral diseases with aptamers.

Pan et al. (1995) selected RNA molecules that neutralize RSV in a concentration-dependent manner. These RNAs were selected by incubation of the random RNA library with whole RSV particles and subsequent separation of the non-virus bound aptamers from those that remained immobilized on the surface of the virus particles. Consequently, the isolated aptamers were shown to interact with the virus surface, unable to enter the virus itself. In a virus-neutralizing activity assay the yield of viral protein could be reduced to 85%–92% at RNA concentrations of 20.0nM. Virus replication was completely blocked at an aptamer concentration of 160.0nM. It was assumed that the presence of the virus-specific aptamers resulted in conformational changes of the glycoprotein structures of viral coat proteins, thereby interfering with the attachment of viruses to the targeted cell and with cell-membrane fusion. Substitution of the pyrimidines with 2′-fluoro-2′-deoxypyrimidine derivatives increased the stability against nucleases but led to a significant decrease in virus neutralization properties.

Various proteins of HIV-1 which are essential for virus replication have been chosen as targets for isolating specific oligonucleotide antagonists. Allen et al. (1995) selected RNA aptamers directed against HIV-1 integrase. Tuerk et al. (1992) and Schneider et al. (1995) isolated RNA and ssDNA aptamers which recognize HIV-1 reverse transcriptase (HIV-1 RT). The isolated aptamers were able to inhibit the DNA polymerase activity of HIV-1 RT, with values of $K_i$ as low as 0.3nM (ssDNA aptamer). In addition, the aptamers selectively discriminated between various RTs, showing high affinity and specificity for HIV-1 RT whereas they did not bind to AMV RT or MMLV RT. Giver et al. (1993) used the HIV-1 rev protein as a target for aptamer selections. Symensma et al. (1996) showed that RNA aptamers, derived against the HIV-1 Rev protein, can mediate Rev function in vivo. This is an impressive example that describes the in vivo functionality of a RNA aptamer.

## 6 Specific Aptamers with High Affinity Inhibit Enzyme Function

One of the first ssDNA aptamers was directed against human α-thrombin, and ssDNA (Bock et al. 1992), RNA (Kubik et al. 1994) as well as 2′ modified RNA aptamers (Latham et al. 1994) are known for human α-thrombin. Bock et al. (1992) isolated ssDNA molecules that bind human thrombin and inhibited blood clotting in vitro. They showed that the thrombin-dependent blood clotting activity was blocked up to 50% at an aptamer concentration of 25nM. Kubik et al. (1994)

selected RNA- and modified RNA aptamers for the same target. For all of these oligonucleotides, mapping of the binding site suggested that they bound to the heparin-binding exo site. The biological role that α-thrombin plays in mechanisms of blood clotting indicates a potential medical application. Indeed, the selected aptamers show inhibitory function of this process in vitro and potential applications as anticoagulants in an ex vivo, whole artery angioplasty model and in preclinical studies with dogs (LI et al. 1994; DEAUDA et al. 1994).

GAL et al. (1998) selected RNA ligands that inhibit activated protein C (APC) function, an essential protein in thrombosis and hemostasis regulation. The activation of protein C is induced by thrombin complexed to thrombomodulin (ESMON and OWEN 1981). GAL et al. showed that the selected aptamer, designated as APC-99, inhibits APC function with a $k_i$ of 137 nM and do not interact with related serine-proteases like thrombin, human neutrophil elastase, or hepatitis C virus NS3 protease.

Human neutrophil elastase (hNE) plays a crucial role in several inflammatory diseases, such as septic shock, myocardial ischemia-reperfusion injury, emphysema, and especially in acute respiratory distress syndrome (ARDS) (DORING 1994; REPINE 1992). Inhibitors for proteases are well known but they generally react unspecifically with the active site (EDWARDS and BERNSTEIN 1994). Valyl phosphonate is an unspecific and irreversible inhibitor for serine proteases (OLEKSYSZYN and POWERS 1991). To aim towards a specific interaction with one particular protease, this small molecule was covalently attached to every pool member of a randomized ssDNA library. This selection method, designated as blended-SELEX (GOLD et al. 1995), was used to isolate specific RNA (SMITH et al. 1995) as well as ssDNA (CHARLTON et al. 1997) inhibitors of hNE. The aptamers that best promoted the covalent reaction of the reactive moiety inactivated elastase with a $k_{obs}$ around $2 \times 10^8 M^{-1}$ per min, almost two orders of magnitude faster than peptide-based inhibitors. In a parallel study, the aptamer inhibitor was also shown to reduce lung injury in a rat alveolitis model (BLESS et al. 1997). By using the $^{99m}$Tc-labeled anti elastase aptamer and, as a control, a $^{99m}$Tc-labeled rat anti-elastase IgG, which is clinically used during in vivo imaging of inflammatory sites, it was shown that the aptamer achieved a significantly higher target-to-background (T/B) ratio in less time than the IgG (Fig. 2).

This example indicates that in some cases specific ligand binding nucleic acids might be advantageous over antibodies; in the present case the superior T/B ratios were attributed to the more rapid clearance of the aptamer from the peripheral circulation compared to the IgG.

High concentrations of human nonpancreatic secretory phospholipase $A_2$ (hnsp-$PLA_2$) are associated with various diseases such as acute pancreatitis, ARDS, bacterial peritonitis, and septic shock (RINTALA and NEVALAINEN 1993). Antagonists with high affinity and specificity are proposed to block the biological role of phospholipase $A_2$ ($PLA_2$). One such antagonist is LY311727, a small molecule indol-based inhibitor of $PLA_2$ (SCHEVITZ et al. 1995). BRIDONNEAU et al. (1998) have chosen a SELEX approach for new drug design and isolated 2′-amino-modified RNA aptamers as $PLA_2$ antagonists. One of these ligands, designated as

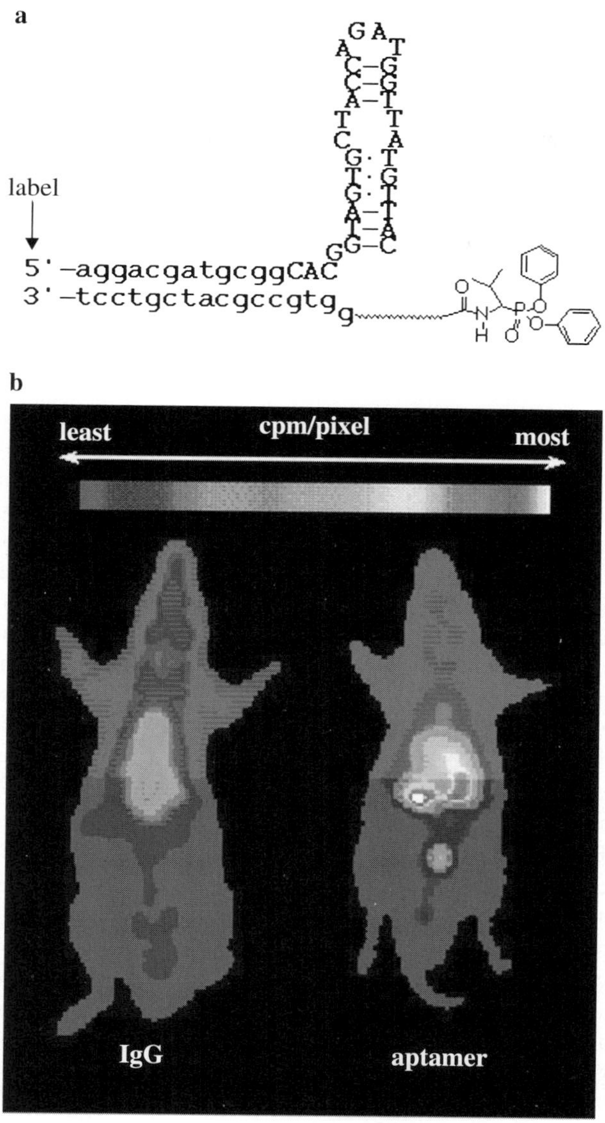


**Fig. 2a,b.** In vivo imaging with an aptamer. **a** DNA aptamer that binds to neutrophil elastase and acts as an irreversible inhibitor via the attached valine derivative. **b** Inflammation imaging by aptamer NX21909 (*right*) and IgG (*left*) after10 min; *cpm*, counts per minute. (From CHARLTON et al. 1997)

aptamer 5, showed a $k_i$ of 0.14nM and 93% inhibition in the chromogenic hnps-$PLA_2$ enzymatic activity assay (REYNOLDS et al. 1992). In tissue-based contraction assays, the truncated version of aptamer 5 showed higher apparent dissociation constants ($k_B = 118 \pm 26$nM) than the non-oligonucleotide based antagonist LY311727 ($270 \pm 50$nM) (SNYDER et al. 1993).

Protein tyrosine phosphatases (PTPases) play important roles in oncogenesis, pathogenesis and the regulation of cellular and viral replication. Common antagonists show little specificity and cannot discriminate between various PTPases. Two different random pools, varying in length, have been exploited to isolate more specific ligands that bind and inhibit the PTPase active site (Bell et al. 1998). The selected aptamers shared a 21-residue conserved sequence independent of the pool they were isolated from and were shown to inhibit the *Yersinia* PTPase used in the selection protocol by $IC_{50}$ values of 10.0 and 35.0nM.

## 7 Aptamers Against Various Target Molecules

An impressive example for the application of combinatorial RNA libraries to identify natural RNA binding sites for a given target protein was provided by Buckanovich and Darnell 1997. They selected RNA aptamers which bind to the neuron-specific RNA binding protein Nova-1, an autoantigen in a neurologic disorder associated with breast cancer and dysfunction of brainstem or spinal motor systems (Buckanovich et al. 1996). Although it was known that Nova-1 contained three RNA binding domains, a particular RNA motif recognized by Nova-1 was unknown. To identify such motifs, an in vitro selection for Nova-1 binders was performed using a library of $10^{15}$ different 52-mer RNAs. Isolated aptamers contained a conserved 15-mer consensus motif $[UCAU(N)_{0-2}]_3$ which was found to be absolutely necessary for Nova-1 binding. Remarkably, a GenBank search for this consensus sequence identified natural Nova-1 binding sites in two neuronal pre-mRNAs. One sequence lies within an intron of the glycine receptor $\alpha 2$ (GlyR $\alpha 2$) pre-mRNA. The other sequence was the pre-mRNA which encodes for Nova-1 itself. Both natural pre-mRNAs specifically interact with authentic Nova-1 protein. These studies established that Nova-1 functions as a sequence-specific nuclear RNA binding protein in vivo. It was one of the first SELEX experiments that identified previously unknown, naturally occurring, nucleic acid binding sites of a RNA binding protein. The authors suggested that the mechanism of the neurologic disease involves disruption of Nova-1 binding to GlyR $\alpha 2$ pre-mRNA by the autoantibody.

Vasopressin is known as an important peptide hormone which regulates water balance in the body (Nielsen et al. 1995). This peptide plays a key role in various states of disease including diabetes insipidus as well as hyponatremia and polydipsia in schizophrenic patients (Goldman et al. 1997). Williams et al. (1997) used the SELEX method to isolate L-ssDNA ligands to vasopressin. In their approach D-DNA ligands have been selected using D-vasopressin as a target molecule. The enantiomer of the winning D-ssDNA aptamer, designated as L-ssDNA aptamer, has been synthesized and its ability to bind L-vasopressin was demonstrated. Importantly, this approach led to enhanced nuclease stability by mirror-image ssDNA (Klussmann et al. 1996; Nolte et al. 1996). The L-ssDNA-aptamer inhibited cAMP release mediated by vasopressin, but the cAMP release induced by oxytocin was not affected.

Human thyroid stimulating hormone (hTSH) belongs to a family of glycohormone proteins such as leutinizing hormone (hLH), follicle stimulating hormone (hFSH), and chorionic-gonadotropin (hCG) (Pierce and Parsons 1981). The measurement of serum hTSH levels, secreted by the pituitary gland, is important for diagnosis of pituitary and thyroid disorders such as hyperthyroidism and hypothyroidism (Labrie 1990). Lin et al. (1996) isolated specific ligands for hTSH that do not interact with the other members of the same protein family. They showed that selected oligonucleotides could be used in diagnostic assays.

Haller and Sarnow (1997) isolated RNA molecules that show a 1000-fold higher affinity for N-7 methylguanosine than for nonmethylated guanosines. A remarkable feature of the selected RNA is that binding to the 5′ terminal cap structure leads to the inhibition of cap-dependent processes such as mRNA translation initiation. The cap structure plays an important role in many different biological processes, e.g., pre-mRNA splicing, nuclear transport, and mRNA stabilization. Haller and Sarnow showed that the isolated aptamer R8-35 not only displaces human cytoplasmic cap-binding complex, but also inhibits the formation of human nuclear cap-binding protein complexes like CBP20 and CBP80. This aptamer is an impressive example for selective discrimination between two highly related molecules with only small structural differences. In the reported case, aptamers provide useful tools to investigate and understand biological processes in living cells.

Another example indicating that remarkably specific aptamers can be isolated by SELEX was reported by Yang et al. (1998), who selected DNA ligands that bind to cellobiose, the disaccharide of cellulose. The selected motifs are able to discriminate sugar epimers, anomers, and other disaccharides. Oligosaccharides play an important role in cellular adhesion, inflammation and molecular recognition. Commonly, antibodies are used for diagnostic purposes but they often cannot discriminate between various sugar tags of cell surface receptors.

## 8 Summary

We have listed and described recent promising developments in the field of aptamer research. The properties and the application potential of aptamers propose an exciting future for aptamers either in the clinic or as research tools for various purposes. We have reviewed exciting examples in which the SELEX technology was applied to obtain promising tools that may help to facilitate our understanding of biological processes and to interfere at distinct points in signal transduction cascades. High affinities and specificities of aptamer/target-interactions can now routinely be achieved. Furthermore, a wide spectrum of chemical modifications of nucleotides is known which greatly increase the stability of RNA molecules in biological materials, considerably enhancing their application potential. The aptamer technology shows that the combination of organic synthesis and molecular biology can contribute to interesting and promising new drug leads, which may very soon find their way into daily clinical practice or onto the laboratory benches of many researchers in the life sciences.

*Acknowledgements.* We thank D. Proske, A. Jenne and M. Blind for critical reading of the manuscript.

## References

Allen P, Worland S, Gold L (1995) Isolation of high-affinity RNA ligands to HIV-1 integrase from a random pool. Virology 209:327–336

Bell SD, Denus JM, Dixon JE, Ellington AD (1998) RNA molecules that bind to and inhibit the active site of a tyrosine phosphatase. J Biol Chem 273:14309–14314

Bevilacqua MP, Nelson RM (1993) Selectins. J Clin Invest 91:379–387

Bevilacqua MP, Nelson RM, Mannori G, Cecconi O (1994) Endothelial-Leukocyte adhesion molecules in human disease. Annu Rev Med 45:361–378

Bless NM, Ward PA (1997) Protective effects of an aptamer inhibitor of neutrophil elastase in lung inflammatory injury. Curr Biol 7:877–880

Bock LC, Griffin LC, Latham JA, Vermaas EH, Toole JJ (1992) Selection of single-stranded DNA molecules that bind and inhibit human thrombin. Nature 355:564–566

Brauchle M, Madlener M, Wagner AD, Angermeyer K, Lauer U, Hofschneider PH (1996) Keratinocyte growth factor is highly overexpressed in inflammatory bowel disease. Am J Pathol 149:521–529

Bridonneau P, Chang YF, O'Connell D, Gill SC, Snyder DW, Johnson L, Goodson TJ, Herron DK, Parma DH (1998) High-affinity aptamers selectively inhibit human nonpancreatic secretory phospholipase $A_2$ (hnsp-$PLA_2$). J Med Chem 41:778–786

Buckanovich RJ, Darnell RB (1997) The neuronal RNA binding protein NOVA-1 recognises specific RNA targets in vitro and in vivo. Mol Cell Biol 17:3194–3201

Buckanovich RJ, Yang YY, Darnell RB (1996) The onconeural antigen Nova-1 is a neuron-specific RNA binding protein, the activity of which is inhibited by paraneoplastic antibodies. J Neurosci 16: 1114–1122

Charlton J, Kirschenheuter GP, Smith D (1997) Highly potent irreversible inhibitors of neutrophil elastase generated by selection from a randomized DNA-valine phosphonate library. Biochemistry 36:3018–3026

Charlton J, Sennello J, Smith D (1997) In vivo imaging of inflammation using an aptamer inhibitor of human neutrophil elastase. Chem Biol 4:809–816

Dagleish AG, Beverly PCL, Clapham PR, Crawford MF, Greaves MF, Weiss RA (1984) The CD4 (T4) antigen is an essential component of the receptor for the AIDS retrovirus. Nature 312:763

DeAnda JA, Coutre SE, Moon MR, Vial CM, Griffin LC, Law VS, Komeda M, Leung LLK, Miller DC (1994) Pilot study of the efficacy of a thrombin inhibitor for use during cardiopulmonary bypass. Ann Thorac Surg 58:344–350

Doring G (1994) The role of neutrophil elastase in chronic inflammation. Am J Respir Crit Care Med 150:114–117

Doudna JA, Cech TR, Sullenger BA (1995) Selection of an RNA molecule that mimics a major autoantigenic epitope of human insulin receptor. Proc Natl Acad Sci USA 92:2355–2359

Eaton BE (1997) The joys of in vitro selection: chemically dressing oligonucleotides to satiate protein targets. Curr Opin Chem Biol 1:10–16

Edwards P, Bernstein P (1994) Synthetic inhibitors of elastase. Med Res Rev 14:127–194

Ellington AD, Szostak JW (1990) In vitro selection of RNA molecules that bind specific ligands. Nature 346:818–822

Esmon CT, Owen WG (1981) Identification of an endothelial cell cofactor for thrombin-catalyzed activation of protein C. Proc Natl Acad Sci USA 78:2249–2252

Farrar MA, Schreiber RD (1993) The molecular cell biology of interferon-$\gamma$ and its receptors. Annu Rev Immunol 11:571

Finch PW, Pricolo V, Wu A, Finkelstein SD (1996) Increased expression of keratinocyte growth factor messenger RNA associated with inflammatory bowel disease. Gastroenterology 110:441–451

Finch PW, Rubin JS, Miki T, Ron D, Aaronson SA (1989) Human KGF is FGF-related with properties of a paracrine effector of epithelial cell growth. Science 244:752–755

Folkman J, Klagsbrun M (1987) Angiogenic factors. Science 235:442–447
Gal SW, Amontov S, Urvil PT, Vishnuvardhan D, Nishikawa F, Kumar PKR, Nishikawa S (1998) Selection of a RNA aptamer that binds to human activated protein C and inhibits its protease function. Eur J Biochem 252:553–562
Giver L, Bartel DP, Zapp ML, Green MR, Ellington AD (1993) Selection and design of high-affinity RNA ligands for HIV-1 Rev. Gene 137:19–24
Gold L, Polisky B, Uhlenbeck O, Yarus M (1995) Diversity of oligonucleotide functions. Annu rev Biochem 64:763–797
Goldman MB, Robertson GL, Luchins DJ, Hedecker D, Pandey GN (1997) Psychotic exacerbations and enhanced vasopressin secretion in schizophrenic patients with hyponatremia and polydipsia. Arch Gen Psychiatry 54:443–449
Gospodarowitz D (1991) Fibroblast growth factors: from genes to clinical applications. Cell Biol Rev 25:307–314
Green LS, Jellinek D, Bell C, Beebe LA, Feistner BD, Gill SC, Jucker FM, Janjic N (1995) Nuclease-resistant nucleic acid ligands to vascular permeability factor/vascular endothelial growth factor. Chemistry & Biology 2:683–695
Green LS, Jellinek D, Jenison R, stman A, Heldin CH, Janjic N (1996) Inhibitory DNA ligands to platelet-derived growth factor B-chain. Biochemistry 35:14413–14424
Haller AA, Sarnow P (1997) In vitro selection of a 7-methyl-guanosine binding RNA that inhibits translation of capped mRNA molecules. Proc Natl Acad Sci USA 94:8521–8526
Iida H, Seifert R, Alpers CE, Gronwald RGK, Phillips PE, Pritzl P, Gordon K, Gown AM, Ross R, Bowen-Pope DF, Johnson RJ (1991) Platelet-derived growth factor (PDGF) and PDGF receptor are induced in mesangial proliferation nephritis in the rat. Proc Natl Acad Sci USA 88:6560–6564
Ishil H, Hattori Y, Itoh H, Kishi T, Yoshida T, Sakamoto H (1994) Preferential expression of the third immunoglobulin-like domain of k-sam product provides keratinocyte growth factor-dependent growth in carcinoma cell lines. Cancer Res 54:518–522
Jellinek D, Green LS, Bell C, Lynott CK, Gill N, Vargeese C, Kirschenheuter G, McGee DP, Abesinghe P, Pieken WA, et al. (1995) Potent 2′-amino-2′-deoxypyrimidine RNA inhibitors of basic fibroblast growth factor. Biochemistry 34:11363–11372
Jellinek D, Lynott CK, Rifkin DB, Janjic N (1993) High-affinity RNA ligands to basic fibroblast growth factor inhibit receptor binding. Proc Natl Acad Sci USA 90:11227–11231
Klußmann S, Nolte A, Bald R, Erdmann VA, Fürste JP (1996) Mirror-image RNA that binds D-adenosine. Nat Biotechnol 14:1112–1115
Kraus E, James W, Barclay AN (1998) Cutting edge: Novel RNA ligands able to bind CD4 antigen and inhibit CD4+ T lymphocyte function. J Immunology 160:5209–5212
Kubik MF, Bell C, Fitzwater T, Watson SR, Tasset DM (1997) Isolation and characterization of 2′-fluoro-, 2′-amino-, and 2′-fluoro-/amino-modified RNA ligands to human IFN-gamma that inhibit receptor binding. J Immunol 159:259–267
Kubik MF, Stephens AW, Schneider D, Marlar RA, Tasset D (1994) High-affinity RNA ligands to human alpha-thrombin. Nucleic Acids Res 22:2619–2626
Labrie F (1990) Hormones from molecules to diseases, E. E. Baulieu and P. A. Kelly, eds. (New York, N.Y.: Herman Publisher).
Latham JA, Johnson R, Toole JJ (1994) The application of a modified nucleotide in aptamer selection: novel thrombin aptamers containing 5-(1-pentynyl)-2′-deoxyuridine. Nucleic Acids Res 22:2817–2822
Lee PP, Ramanathan M, Hunt CA, Garovoy MR (1996) An oligonucleotide blocks interferon-gamma signal transduction. Transplantation 62:1297–1301
Lee SW, Sullenger BA (1997) Isolation of a nuclease-resistant decoy RNA that can protect human acetylcholine receptors from myasthenic antibodies. Nature Biotechnology 15:41–45
Li WX, Kaplan AV, Grant GW, Toole JJ, Leung LLK (1994) A novel nucleotide-based thrombin inhibitor inhibits clot-bound thrombin and reduces arterial platelet thrombus formation. Blood 83:677–682
Lin Y, Nieuwlandt D, Magallanez A, Feistner B, Jayasena SD (1996) High-affinity and specific recognition of human thyroid stimulating hormone (hTSH) by in vitro selected 2′-amino-modified RNA. Nucleic Acid Research 24:3407–3414
Lindner V, Giachelli CM, Schwartz SM, Reidy MA (1995) A subpopulation of smooth muscle cells in injured rat arteries expresses platelet-derived growth factor-B chain mRNA. Circ Res 76:951–957
Lindner V, Reidy MA (1995) Platelet-Derived growth factor ligand and receptor expression by large vessel endothelium in vivo. Am J Pathol 146:1488–1497
Lindstrom J, Shelton D, Fujii Y (1988) Myasthenia gravis. Advances in Immunology 42:233–284

Moscatelli D (1987) High and low affinity sites from basic fibroblast growth factor on cultured cells: absence of a role for low affinity binding in the stimulation of plasminogen activator production by bovine capillary endothelial cells. J Cell Physiol 131:123–130

Nielsen S, Chou CL, Marples D, Christensen EI, Kishore BK, Knepper MA (1995) Vasopressin increases water permeability of kidney collecting duct by inducing translocation of aquaporin-CD water channels to plasma membrane. Proc Natl Acad Sci USA 92:1013–1017

Nolte A, Klußmann S, Bald R, Erdmann VA, Fürste JP (1996) Mirror-design of L-oligonucleotide ligands binding to L-arginine. Nat Biotechnol 14:1116–1119

O'Connell D, Koenig A, Jennings S, Hicke B, Han HL, Fitzwater T, Chang YF, Varki N, Parma D, Varki A (1996) Calcium-dependent oligonucleotide antagonists specific for L-selectin. Proc Natl Acad Sci USA 93:5883–5887

Oleksyszyn J, Powers J (1991) Irreversible inhibition of serine proteases by peptide derivatives of (alpha-aminoalkylphosphonate)diphenyl ester. Biochemistry 30:485–493

Osborne SE, Ellington AD (1997) Nucleic acid selection and the challenge of combinatorial chemistry. Chem Rev 97:349–370

Pagratis NC, Bell C, Chang Y-F, Jennings S, Fitzwater T, Jellinek D, Dang C (1997) Potent 2′-amino-, and 2′-fluoro-2′-deoxyribonucleotide RNA inhibitors of keratinocyte growth factor. Nature Biotechnology 15:68–73

Pan W, Craven RC, Qiu Q, Wilson CB, Wills JW, Golovine S, Wang JF (1995) Isolation of virus-neutralizing RNAs from a large pool of random sequences. Proc Natl Acad Sci USA 92:11509–11513

Pierce JG, Parsons IF (1981) Glycoprotein hormones: structure and function. Annu Rev Biochem 50:465–495

Repine J (1992) Scientific perspectives on adult respiratory distress syndrome. Lancet 339:466–469

Reynolds RJ, Hughes LL, Denis EA (1992) Analysis of human synovial fluid phospholipase A2 on short chain phosphatidylcholine-mixed micelles: development of a spectrophotometric assay suitable for a microtiterplate reader. Anal Biochem 204:190–197

Rintala EM, Nevalainen TJ (1993) Group II phospholipase A2 in sera of febrile patients with microbiologically or clinically documented infections. Clin Infect Disease 17:864–870

Rubin JS, Osada H, Finch RW, Taylor WG, Rudikoff S, Aaronson SA (1989) Purification and characterization of a newly identified growth factor specific for epithelial cells. Proc Natl Acad Sci USA 86:802–806

Schevitz RW, Bach NJ, Carlson DG, Chirgadze NY, Clawson DK, Dillard RD, Draheim SE, Hartley LW, Jones ND, Mihelich ED, Olkowski JL, Snyder DW, Sommers C, Wery JP (1995) Structure-based design of the first potent and selective inhibitor of human non-pancreatic secretory phospholipase A2. Nature Struct Biol 2:458–465

Schneider DJ, Feigon J, Hostomsky Z, Gold L (1995) High-affinity ssDNA inhibitors of the reverse transcriptase of type 1 human immunodeficiency virus. Biochemistry 34:9599–9610

Smith D, Kirschenheuter GP, Charlton J, Guidot DM, Repine JE (1995) In vitro selection of RNA-based irreversible inhibitors of human neutrophil elastase. Chemistry & Biology 2:741–750

Snyder DW, Sommers CD, Bobbitt JL, Mihelich ED (1993) characterization of the contractile effects of human recombinant non-pancreatic secretory phospholipase A2 and other PLA2 s on guinea pig lung pleural-strips. J Pharmacol Exp Ther 266:1147–1155

Staiano-Coico L, Krueger JG, Rubin JS, D'limi S, Vallat VP, Valentino L (1993) Human keratinocyte growth factor effect in a porcine model of epidermal wound healing. J Ex Med 178:865–878

Symensma TL, Giver L, Zapp M, Takle GB, Ellington AD (1996) RNA aptamers selected to bind human immunodeficiency virus type 1 Rev in vitro are Rev responsive in vivo. J Virol 70:179–187

Trinchieri G, Perussia B (1985) Immune interferon: a pleiotropic lymphokine with multiple effects. Immunology Today 6:131

Tuerk C, Gold L (1990) Systematic evolution of ligands by exponential enrichment: RNA ligands to bacteriophage T4 DNA polymerase. Science 249:505–510

Tuerk C, MacDougal S, Gold L (1992) RNA pseudoknots that inhibit human immunodeficiency virus type 1 reverse transcriptase. Proc Natl Acad Sci USA 89:6988–6992

Wiegand TW, Williams PB, Dreskin SC, Jouvin MH, Kinet JP, Tasset D (1996) High-affinity oligonucleotide ligands to human IgE inhibit binding to Fc epsilon receptor I. J Immunol 157:221–230

Williams KP, Liu XH, Schumacher TN, Lin HY, Ausiello DA, Kim PS, Bartel DP (1997) Bioactive and nuclease-resistant L-DNA ligand of vasopressin. Proc Natl Acad Sci USA 94:11285–11290

Yang Q, Goldstein IJ, Mei HY, Engelke DR (1998) DNA ligands that bind tightly and selectively to cellobiose. Proc Natl Acad Sci USA 95:5462–5467

# In Vitro Selection of Nucleic Acid Enzymes

M. KURZ[2] and R.R. BREAKER[1]



## 1 Introduction

Among the enormous quantity and diversity of enzymes found in nature, researchers have confirmed the existence of only seven distinct classes of biological catalysts that are made of RNA rather than protein. As currently understood, the natural functions of these ribozymes are limited to phosphoester transfer and phosphoester hydrolysis reactions that occur with RNA or DNA substrates (KRUGER et al. 1982; GUERRIER-TAKADA et al. 1983; PEEBLES et al. 1986; PRODY et al. 1986; BUZAYAN et al. 1986; SHARMEEN et al. 1988; SAVILLE and COLLINS 1990; ZIMMERLY et al. 1995). Although ribozymes are exceedingly rare and their biochemical functions are limited, they serve as essential components of the metabolic

[1] Department of Molecular, Cellular and Developmental Biology, PO Box 208103, Yale University, New Haven, CT 06520-8103, USA
[2] *Current address*: Phylos Inc., 128 Spring Street, Lexington, MA 02421, USA

machinery of all living systems. It has been proposed that nucleic acids preceded proteins in the evolutionary history of biocatalysis, and that these primitive ribozymes catalyzed many of the reactions performed by modern protein enzymes (Gilbert 1986; Benner et al. 1989; Hirao and Ellington 1995). Over the course of 4 billion years of evolution, it appears that nature has determined that proteins are a superior format for constructing enzymes; however, the true capacity of nucleic acids for catalytic function remains to be fully defined.

Like all other organic chemists, nature must abide by the same set of immutable chemical principles as the process of evolution runs its course. Considering the expanded chemical composition of proteins vs nucleic acids, it would be difficult to contest the results of billions of years of evolution by arguing that the enzymatic potential of these two very different polymers are equivalent. However, it might be possible to harness the catalytic potential of nucleic acids in the form of artificial engineered enzymes that display useful catalytic functions. In recent years, the narrow repertoire of chemical reactions that are catalyzed by existing ribozymes has undergone dramatic expansion through the use of 'in vitro selection', a process that allows enzyme engineers to rapidly 'evolve' new ribozymes outside the confines of living cells (Williams and Bartel 1996; Breaker 1997a). This breakthrough in enzyme engineering technology has made it possible for researchers to access the true catalytic potential of nucleic acids and to contemplate practical applications for novel 'designer' ribozymes.

Many of the ribozymes created to date using in vitro selection methods display catalytic functions that are similar to those seen with natural protein enzymes. For example, a variety of new ribozymes has been created that catalyze reactions at the phosphorus centers within nucleic acid substrates, as is typically seen with natural ribozymes. Among this expanding collection of novel ribozymes are RNAs that catalyze RNA cleavage (Pan and Uhlenbeck 1992; Williams et al. 1995), DNA cleavage (Beaudry and Joyce 1992), RNA ligation (Bartel and Szostak 1993; Hager and Szostak 1997), RNA polymerization (Ekland and Bartel 1996), phosphoryl coupling (Chapman and Szostak 1995; Huang and Yarus 1997a,b) and phosphoryl transfer (Lorsch and Szostak 1994). Ribozymes also have been made to promote carboxylate ester bond formation (Illangasekare et al. 1995; Lohse and Szostak 1996; Jenne and Famulok 1998), cleave (Dai et al. 1995, 1996) and form amide bonds (Lohse and Szostak 1996; Zhang and Cech 1997; Wiegand et al. 1997), and metalate porphyrin rings (Conn et al. 1996). These examples show that many of the reactions of modern biochemistry are accessible to ribozymes.

Even chemical transformations that one would more likely find in a chemist's flask rather than the cytoplasm of a cell can be performed by artificial ribozymes (Frauendorf and Jaschke 1998). These reactions include alkylation by halide displacement (Wilson and Szostak 1995; Wecker et al. 1996), isomerization of a bridged biphenyl compound (Prudent et al. 1994), and Diels-Alder reactions (Tarasow et al. 1997). Without question, the catalytic functions of nucleic acids will continue to be diversified through the use of in vitro selection and other enzyme engineering strategies. This review will highlight the recent progress made in the area of ribozyme design, and report on the current state of technical capability in the field.

# 2 New Unmodified Ribozymes

## 2.1 Amide and Peptide Bond Synthesis

It is speculated that RNA played a central role in early biochemical development, and this predecessor metabolic state that was maintained by early ribozymes eventually gave rise to encoded protein synthesis and soon thereafter to modern protein enzymes. If this view is correct, then a ribozyme that could catalyze amide bond formation, must have emerged to make this transition possible. This 'amide synthase' activity is a significant departure from the catalytic activities of the seven natural classes of ribozymes. However, it is intriguing that ribosomal RNA which has been stripped of nearly all proteins retains significant peptidyl transferase activity (NOLLER et al. 1992), suggesting that the modern cellular machinery that is responsible for all protein synthesis indeed may be an amide synthase ribozyme.

Following the initial discovery of ribozymes and the expansion of the RNA world hypothesis, many endeavors have been undertaken to understand the overall scope of RNA catalysis and to bridge the gap from 'primitive' RNA catalysis to the advanced enzyme function that we find in the modern protein world (HAGER et al. 1996). To test whether RNA is capable of forming amide and peptide bonds, several groups have employed in vitro selection to seek artificial enzymes with these activities. LOHSE and SZOSTAK (1996) provided the first evidence that amide bond formation is part of the catalytic repertoire of RNA (Fig. 1a; SUGA et al. 1998). Two more recent discoveries provide further evidence that RNA has the ability to participate in protein synthesis. In the first example (WIEGAND et al. 1997; see also Sect. 4), used selection to isolate chemically modified RNA enzymes that promote the attack of a primary amine on an AMP-carboxylic acid mixed anhydride, which results in the formation of an amide linkage (Fig. 1b). Using a similar strategy, ZHANG and CECH (1997) obtained ribozymes that could couple a modified methionine-AMP-ester to a tethered phenylalanine residue in a reaction that closely resembles the chemical process of the natural peptidyl transferase reaction (Fig. 1c). The modified methionine-AMP-ester, which serves as a model for an aminoacylated tRNA substrate, is bound to an artificial 'P site'. Here it is subject to the nucleophilic attack of the phenylalanine amino group bound to an artificial 'A site', leading to the formation of a true peptide bond. Some ribozyme variants with this activity achieve observed rate constants near 0.1/min, corresponding to a rate enhancement of at least $10^5$-fold over the uncatalyzed rate. Substrate recognition is mainly governed by the adenosine moiety of the activated amino acid, allowing amino acids other than methionine to be transferred. This characteristic is a prerequisite for a generalized protein synthesizing ribozyme, similar to the ribosome's ability to make use of the many different tRNAs that are individually charged with natural amino acids.

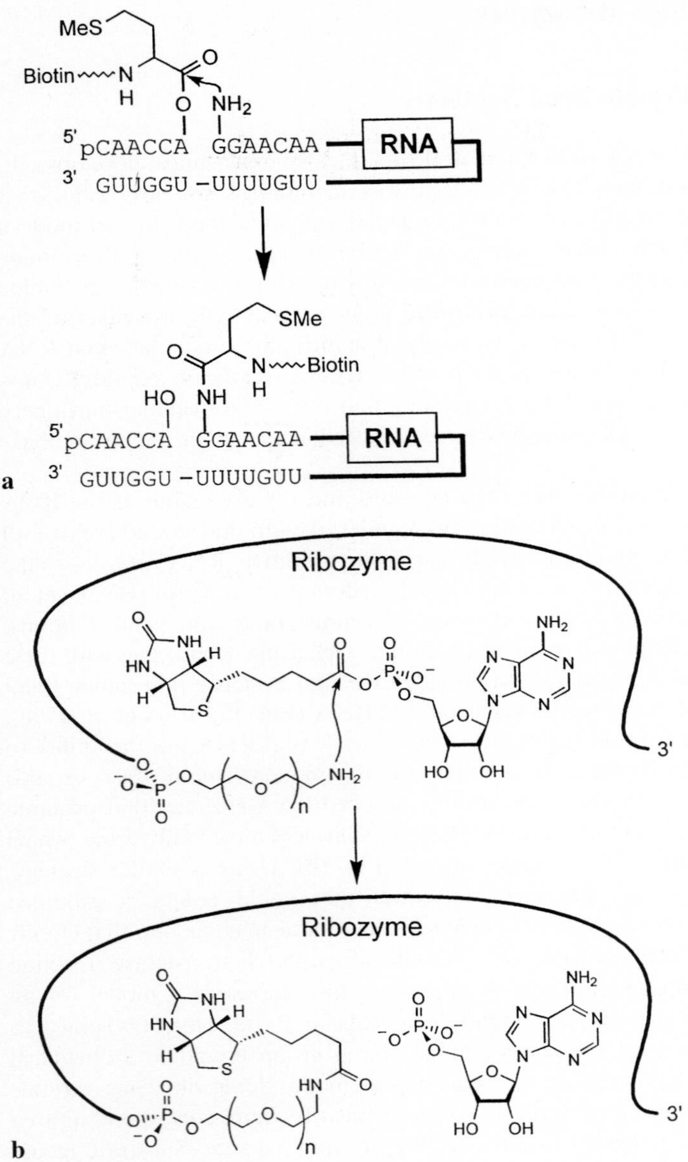


**Fig. 1a–c.** Mechanisms for amide-bond formation by ribozymes reported by **a** LOHSE and SZOSTAK (1996), **b** WIEGAND et al. (1997), and **c** ZHANG and CECH (1997). The mechanism in **a** also operates with a 5′ hydroxyl nucleophile in place of the 5′-amine group as depicted, thereby resulting in acylation of the ribozyme. In **b**, the ribozyme carries imidazole-modified uridine residues that are required for function. In **c**, the *A site* and *P site* are adapted from the nomenclature of the ribosome to relate the similar functions of the natural peptidyl transferase enzyme to this novel ribozyme

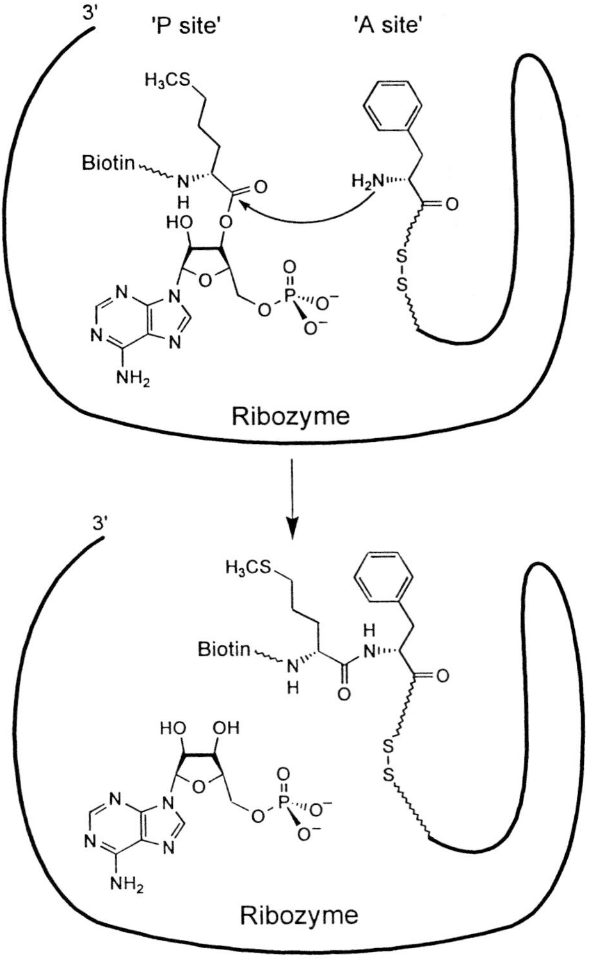


c

**Fig. 1.**

## 2.2 Carbon Ester Bond Formation

Using a closely related approach to that described above, JENNE and FAMULOK (1998) did not find ribozymes with the desired peptidyl transferase activity, but instead isolated ribozymes that catalyzed acyl transfer of the amino acid to an internal ribose-2′-hydroxyl. This reaction is inhibited by AMP not only by competitive binding to the active site, but also by transferring the 2′-acyl group back to AMP in the reverse reaction, with an equilibrium constant which lies strongly on the side of the aminoacyl-AMP starting material (Fig. 2). The isolation of acyl-transferase ribozymes as opposed to peptidyl transferase ribozymes in this selection is not surprising, considering the fact that individual RNAs comprising the selec-

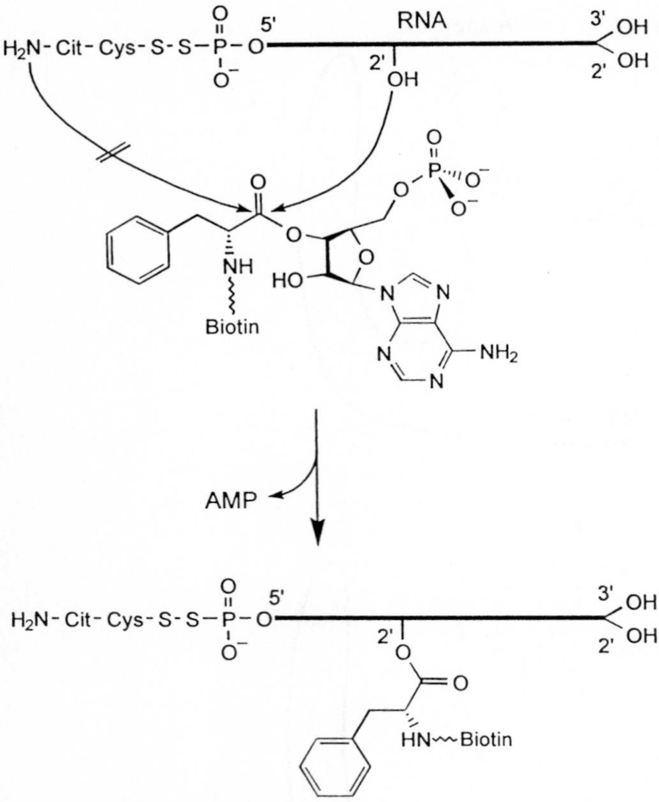


**Fig. 2.** Ester bond formation by a ribozyme. Coupling of the biotinylated phenylalanine compound with the RNA construct could have proceeded by nucleophilic attack of the 5′-amino group, or similarly via attack by one of the many ribose hydroxyl groups. One class of RNAs that populate the pool after 13 rounds of selection displays acyl transferase activity, and generates a covalent ribozyme product wherein the biotinylated amino acid is coupled to a specific internal 2′ hydroxyl. *Cit* and *Cys* represent citrulline and cysteine, respectively

tion pool each carry 159 2′-hydroxyl groups, whereas the desired amine nucleophile is represented only once per molecule. In addition, efficient ribozymes with acyl transferase activity have been isolated previously (LOHSE and SZOSTAK 1996; Fig. 1a). This is just one example of many in vitro selection experiments in which the final outcome was different from the expected outcome (JOYCE 1992; BREAKER 1997a; see Sect. 6).

## 2.3 Phosphoanhydride Exchange and Hydrolysis Reactions

Another example of the inherent independence and inventiveness of the in vitro selection process is evident in a series of publications by HUANG and YARUS

(1997a–c). The authors devised a selection protocol that was intended to yield RNAs that catalyze the nucleophilic attack of an amino acid oxyanion on an RNA triphosphate, thereby forming the corresponding mixed anhydride with concomitant release of inorganic pyrophosphate. Amino acids are similarly activated by aminoacyl tRNA synthetases as part of the natural process of charging tRNAs. Demonstrating this ribozyme activity therefore would provide more complete evidence that RNA is capable of catalyzing the entire series of reactions that are necessary to emulate modern encoded protein synthesis. The ribozymes that resulted from the selection indeed could release pyrophosphate; however this activity proceeds even in the absence of amino acids (HUANG and YARUS 1997a). Further investigation revealed that these RNAs catalyze a phosphate-phosphate anhydride exchange reaction (Fig. 3) with a variety of substrate molecules that possess a phosphate moiety, or they catalyze the competing phosphoanhydride hydrolysis reaction via attack by water (HUANG and YARUS 1997b,c). Other possible reactants include nucleotides, nucleotide-like cofactors such as CoA, NADP, thiamine phosphate, FMN, phosphorylated RNAs, and even phosphorylated non-nucleotide molecules.

The ribozymes isolated by Huang and Yarus (see also CHAPMAN and SZOSTAK 1995) produce nucleoside-5′,5′-pyrophosphate 'caps' that are identical to the chemical structures that are important for cellular RNA processing or that serve as

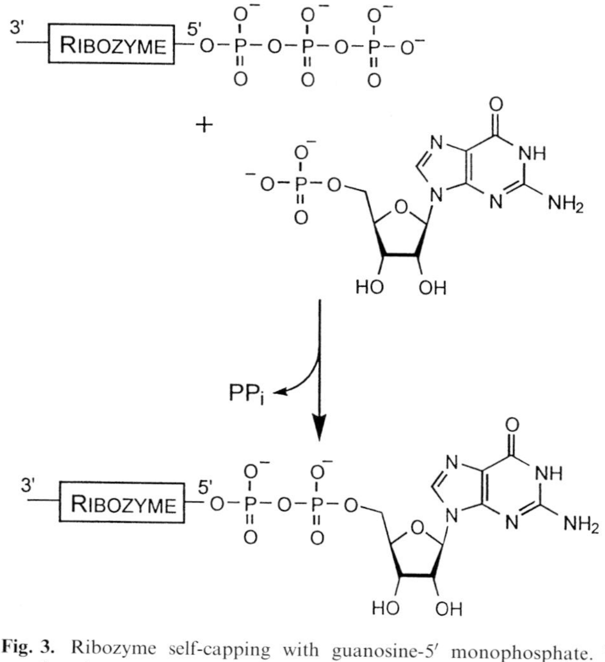


**Fig. 3.** Ribozyme self-capping with guanosine-5′ monophosphate. This phosphoanydride exchange reaction also can proceed with a variety of phosphorylated substrates, or the attacking nucleophile can be replaced by water which results in hydrolysis of the ribozyme 5′-triphosphate moiety. $PP_i$ represents inorganic pyrophosphate

intermediates of enzymatic ligation. In the latter case, protein enzymes such as DNA ligase and RNA ligase activate the 5′ phosphate of the 'donor' oligonucleotide by similarly coupling AMP (using ATP) to generate a 5′,5′-pyrophosphate cap. This intermediate is then attacked by the 3′ oxygen of the corresponding 'acceptor' oligonucleotide to form a new 3′,5′-phosphodiester bond between the two oligonucleotides. HAGER and SZOSTAK (1997) successfully used in vitro selection to isolate ribozymes that could complete the last step of this ligation process (Fig. 4). Presumably, the ribozyme facilitates the attack of the 3′-oxygen of a templated RNA on the 5′,5′-pyrosphosphate linkage of the ribozyme, leading to the formation of a new 3′,5′-phosphodiester bond with concomitant release of AMP. The use of such capped structures as substrates for both modern protein enzymes and ribozymes represent another plausible link between a hypothetical RNA world and the contemporary biochemistry of extant life forms.

## 3 DNA Enzymes

The many examples of natural and artificial ribozymes provide a glimpse at the structural and functional characteristics of RNA and hint at the true catalytic potential for nucleic acids. It is now well accepted that RNA can perform many catalytic functions, but can DNA similarly serve as a catalytic polynucleotide despite the absence of the important 2′-hydroxyl group? For many years after the original discovery of ribozymes, some believed that the 2′-hydroxyl group was key to the formation of RNA catalysts and that DNA is rendered functionally inert by the loss of this group. Clearly, the presence of the 2′-hydroxyl group of every sugar moiety in most instances must be considered advantageous, due to the chemical reactivity of this group and its role in RNA tertiary-structure formation. Despite the inherent limitations that the loss of the 2′-hydroxyl brings about, the catalytic potential for DNA in its single-stranded form may be substantial (BREAKER 1997c).

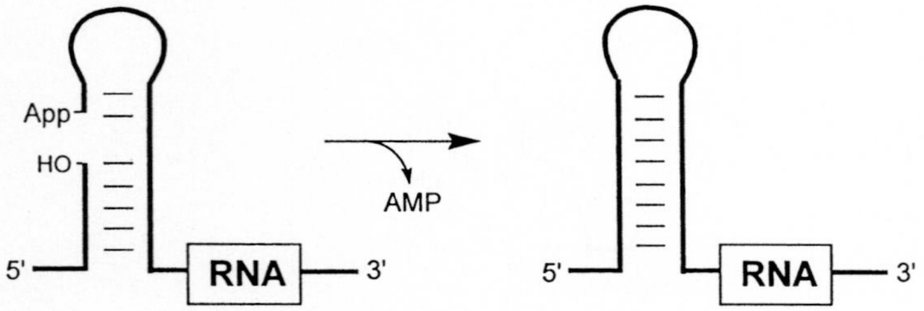


**Fig. 4.** A ribozyme with RNA ligase activity. The 5′,5′-pyrophosphate linkage was incorporated at the 5′ terminus of the ribozyme by in vitro transcription in the presence of excess 5′,5′-AppG. RNAs that promoted RNA ligation via nucleophilic attack by the substrate 3′-OH and release of 5′-AMP are isolated by affinity chromatography and selectively amplified

Although no deoxyribozymes are known to exist in nature, a variety of DNAs that perform enzyme-like functions in vitro have been isolated using selection methods (Breaker and Joyce 1994b, 1995; Couenoud and Szostak 1995; Breaker 1997b,c). More recently, many new catalytic DNAs have been reported, including deoxyribozymes that cleave phosphoramidate bonds (Burmeister et al. 1997), that use amino acid cofactors (Roth and Breaker 1998; see Sect. 4) and that metalate porphyrin rings (Li and Sen 1996). In the latter case, the deoxyribozyme adopts a 'G-quartet' structure and binds the porphyrin substrate in a pocket that favors insertion of a metal ion into the ring. The comparison of spectroscopic properties of porphyrin derivatives in their enzyme-bound and free states has allowed the assignment of the substrate's transition state conformation (Li and Sen 1998). The porphyrin is placed in an environment that distorts the ring out of its more favorable planar geometry, thereby increasing the basicity of the porphyrin by 3–4 $_{p}K_{a}$ units, which favors metal ion insertion. A similar transition-state stabilization mechanism may be used for an artificial ribozyme (Conn et al. 1996) and a catalytic antibody (Cochran and Schultz 1990), both of which display similar enzymatic activities.

An important question to address is whether DNA can achieve rate enhancements that are similar to those of natural ribozymes. Santoro and Joyce (1997) reported the isolation of a surprisingly small deoxyribozyme (Fig. 5) that efficiently cleaves RNA substrates under simulated physiological conditions and with multiple turnover. This molecule can be tailored to cleave a variety of RNA substrates and can function with target molecules that mimic HIV-1 mRNAs. The deoxyribozyme displays maximum catalytic activity when cleaving at a purine-pyrimidine junction. However, both the catalytic rates and generality of the substrate specificities of this deoxyribozyme compare favorably with those of natural self-cleaving ribozymes (Symons 1992), indicating that catalytic DNAs can achieve levels of function that are similar to those of biologically relevant ribozymes. The continued development of RNA-cleaving deoxyribozymes with catalytic speeds that match those of ribozymes could open a broad range of potential applications for artificial DNA enzymes in therapeutics and in biotechnology.

Recent studies also have made clear the fact that DNA can form structures that are remarkably intricate (Breaker 1997b). An example of a nucleic acid structure that previously had not been found in any catalytic RNA or DNA was recently found to play a key role in the function of a $Cu^{2+}$-dependent self-cleaving

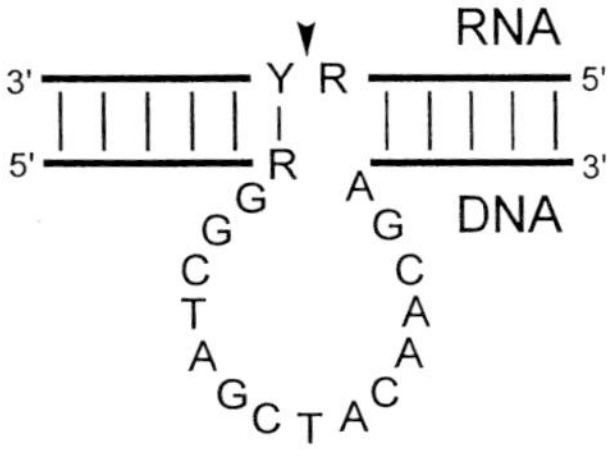


**Fig. 5.** Sequence and secondary structure of an RNA-cleaving deoxyribozyme that functions under simulated physiological conditions. Bases designated comprise the putative catalytic core of the deoxyribozyme. *Y* and *R* represent pyrimidine and purine bases, respectively. The *arrowhead* identifies the site of cleavage by the $Mg^{2+}$-dependent phosphoester transfer reaction

deoxyribozyme isolated by CARMI et al. (1996). This deoxyribozyme catalyzes the region-specific cleavage of DNA substrates via a $Cu^{2+}$-dependent oxidative mechanism. As with the RNA-cleaving DNA described above, this deoxyribozyme can be engineered to cleave separate DNA substrate molecules in a sequence-directed fashion (CARMI et al. 1998). Recognition of separate substrate DNAs by a truncated form of the deoxyribozyme involves duplex and triplex recognition elements, wherein both elements contribute towards defining substrate specificity (Fig. 6). One of the substrate recognition domains of this enzyme binds substrate by Watson/Crick base pairing, while the other arm folds into a short hairpin which binds the target sequence in the context of a DNA triplex through Hoogsteen base pairing. It is likely that the use of triplexes and other higher-ordered structures may provide sufficient structural sophistication for DNA to allow robust catalytic function. If true, then there will be many opportunities in the future to examine new examples of this unlikely catalytic polymer.

# 4 New Cofactors and Reaction Conditions for Nucleic Acid Enzymes

## 4.1 Ribozyme and Deoxyribozyme Catalysis Without Cofactors

Although the scope of catalytic function by RNA and DNA is proving to be extensive, can nucleic acid catalysts compare favorably with the efficiency of protein-based enzymes? Polymers of RNA and DNA are likely to be at a significant disadvantage when competing against polypeptides. Oligonucleotides made from the four standard nucleotides have far less chemical and structural diversity than proteins, which are comprised of 20 different amino acids. In particular, the lack of functional groups that can act as nucleophiles, general acids or general bases is expected to significantly restrict the catalytic potential of nucleic acid enzymes in comparison to protein enzymes. One strategy that could be exploited to improve the chemical and functional diversity of RNA and DNA is the use of external

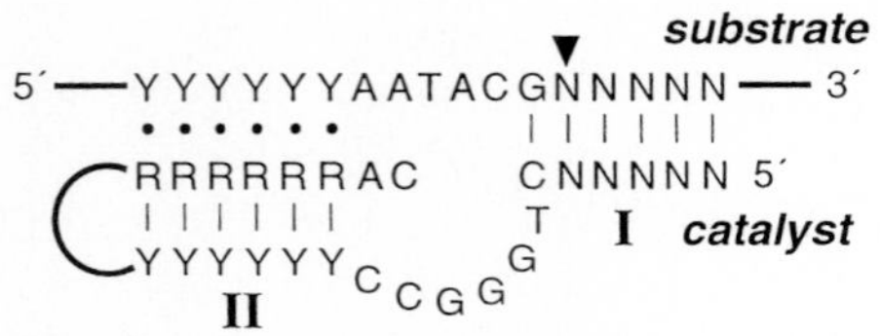


**Fig. 6.** Sequence and secondary structure of a DNA-cleaving deoxyribozyme. Target specificity for this $Cu^{2+}$-dependent DNA enzyme is defined by two substrate-binding domains composed of stems I and II. Stem II forms a DNA triplex interaction wherein the polypyrimidine region of the substrate serves as the 'third strand' of the complex. *N* represents any nucleotide. *Y* and *R* represent pyrimidine and purine nucleotides, respectively. The *arrowhead* identifies the primary site of DNA cleavage

cofactors. Most natural ribozymes discovered to date make use of divalent metal ion cofactors to maximize rate enhancements (Pyle 1993; Yarus 1993). Metal ion cofactors can contribute to ribozyme function by facilitating RNA folding, or they can perform a more important role by serving as necessary components of catalytic sites. Although divalent metal ions can and do play key supportive roles for the action of most natural and artificial nucleic acid catalysts, it is becoming increasingly clear that metal ions are not indispensable for the function of certain ribozymes and deoxyribozymes. For example, several '$Mg^{2+}$-dependent' self-cleaving ribozymes have been known to operate with considerable rate enhancements even when assayed in the complete absence of divalent metals (Dahm and Uhlenbeck 1991; Chowrira et al. 1993; Kuimelis and McLaughlin 1995). More recently, a series of reports confirm that the 'hairpin' self-cleaving ribozyme can achieve full activity without $Mg^{2+}$ if care is taken to stabilize the active tertiary structure of the RNA (Hampel and Cowan 1997; Nesbitt et al. 1997).

Can it be true that nucleic acids, long considered to be entirely reliant on metal ion cofactors for catalytic function, also can form effective active sites using only the functional groups inherent to the polymer or that can be borrowed from other organic compounds? The results of several recent in vitro selection experiments have been used to directly challenge the notion that divalent metal ions are absolutely necessary for the catalytic function of ribozymes and deoxyribozymes. Independently, several research groups (Faulhammer and Famulok 1997; Geyer and Sen 1997; Roth and Breaker 1998) have reported the isolation of deoxyribozymes that cleave RNA phosphoester linkages in the absence of divalent metal ions or any other cofactors. The rate constants achieved by the earliest versions of these artificial deoxyribozymes are only about 100-fold less than those observed for the natural self-cleaving ribozymes, corresponding to a rate enhancement of nearly 1 million-fold over the uncatalyzed rate. It has yet to be determined precisely how the DNA achieves this magnitude of rate enhancement without the use of cofactors. Presumably, DNA can easily position the target RNA linkage for precise 'in-line attack' by the appropriate 2′-hydroxyl group. In addition, it is conceivable that the DNA also might shift the $_{p}K_{a}$ value of a well-placed nucleoside moiety which then can serve as a general acid/base or that can stabilize the transition state of the reaction.

An example of a cofactor-independent ribozyme was isolated under acidic reaction conditions (Jayasena and Gold 1997). An individual self-cleaving ribozyme isolated from a random-sequence pool by in vitro selection displays a maximum rate of catalysis at pH 4.2 in the absence of divalent metals, a finding that adds a new dimension to the potential for ribozyme catalysis. This pH optimum lies close to the $_{p}K_{a}$ values of N1 of the adenosine and N3 of cytidine. In their protonated forms, these bases might be able to serve as general acid catalysts and relieve the need for metal ion cofactors, although the precise catalytic mechanism has yet to be determined. Conventional wisdom would hold that extremes of pH, temperature, or other reaction parameters would prevent the catalytic function of nucleic acid enzymes. However, this report highlights the fact that in vitro selection can be used to produce new RNA and DNA enzymes that operate under reaction conditions that are far from those typically encountered inside cells.

## 4.2 Small Organic Cofactors for Catalytic Polynucleotides

Metal ions are not the only cofactors that can be used by nucleic acids. A unique class of cofactor-dependent deoxyribozymes that use an amino acid to support catalytic activity was isolated recently (Roth and Breaker 1998). This DNA promotes the cleavage of a target RNA linkage only in the presence of L-histidine or closely related analogues. The DNA enzyme presumably binds the histidine cofactor and positions its imidazole moiety for use as a general base, thereby activating the ribose 2′-OH nucleophile by deprotonation (Fig. 7). This mode of operation shows a remarkable similarity to the classical mechanism proposed for ribonuclease A (Walsh 1979). The natural RNA-cleaving protein enzyme also uses a histidine residue at its active site to provide general base catalysis for the first step of the RNA cleavage reaction. This finding suggests that perhaps all of the natural amino acids and many other small organic cofactors may be used by nucleic acid enzymes to increase their limited chemical diversity. Assuming that the structural diversity of nucleic acids is sufficient to bind and precisely position a variety of organic cofactors, then any chemical reaction that can be promoted by protein-based enzymes is a candidate for catalysis by an undiscovered ribozyme or deoxyribozyme.

From the studies described above, we can conclude that both RNA and DNA can actively contribute to the catalytic process rather than simply serving as passive scaffolds for cofactor binding. Nucleic acid enzymes do not need divalent metal ions or any other cofactor to produce catalytic rate enhancements that are relevant at least to biological systems. However, when given the option most in vitro selection experiments yield ribozymes or deoxyribozymes that make use of external cofactors (e.g. Breaker and Joyce 1994b, 1995), indicating that nucleic acid enzymes that are assisted by cofactors are more prevalent than cofactor-independent motifs among the vast number of structural variations that are possible.

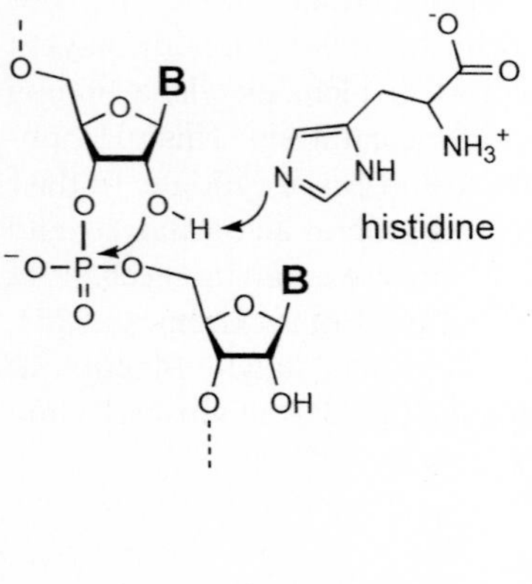


**Fig. 7.** Proposed mechanism of a histidine-dependent DNA enzyme that cleaves RNA

# 5 Nucleic Acid Enzymes with Chemical Modifications

## 5.1 Ribozymes Constructed with Modified Bases

Another approach to overcoming the meager functional diversity of the nucleotide building blocks involves the use of chemically modified RNAs and DNAs. One strategy introduces additional functional groups at defined positions within pre-existing ribozymes or aptamers (Usman et al. 1996; Eaton 1997; Osborne et al. 1997). Here, the primary aims are to increase polymer stability by enhancing nuclease resistance (Lin et al. 1994) or to stabilize secondary structures through connecting oligonucleotide domains with artificial linkers (Osborne et al. 1996; Letsinger and Wu 1995; Macaya et al. 1995; Nelson et al. 1996). An alternative strategy makes use of pre-modified nucleotides to construct random-sequence pools, thereby creating new structural and chemical possibilities that can be screened for functional molecules. In order for this procedure to be successful, non-natural nucleotide triphosphates must be suitable substrates for the reverse transcriptase and RNA polymerase enzymes needed for amplification of the selected RNA or DNA molecules. The list of suitable nucleotide building blocks already is substantial, and includes a large number of structurally diverse 5-substituted pyrimidine and 8-substituted purine bases, as well as nucleotide analogs that carry 2′-sugar modifications. This allows the introduction of virtually any chemical moiety into nucleic acids, such as hydrophobic or hydrophilic groups, acids, bases, nucleophiles and various cross-linkers (Eaton 1997).

Wiegand et al. (1997) provide the first example for the isolation of a modified RNA catalyst isolated from a random pool of modified RNA. In this case, 5-imidazolyl-modified uridine residues were used in place of uridine. This modification is expected to enhance the catalytic power of RNA by conferring added potential for general acid/base catalysis and metal coordination sites in a manner that will not disrupt Watson/Crick base pairing. A collection of amide synthase ribozymes (Fig. 1a), each carrying a short (13 nucleotide) conserved sequence domain, was isolated from the chemically modified RNA pool. The most efficient amide synthase variant displays a $k_{obs}$ of ~0.04/min, corresponding to a rate enhancement of ~$10^5$-fold over the uncatalyzed reaction. This activity is entirely dependent on the presence of the 5-imidazolyl-uridines, confirming that the modifications either play an active role in catalysis or that their removal disrupts the active conformation of the RNA. This and similar chemical modifications are expected to enhance the catalytic ability of RNA. This view receives support from an observation by Wiegand et al. (1997) that a related in vitro selection experiment designed to isolate ester synthases produced significantly slower ribozymes when using unmodified RNA as opposed to beginning with an RNA pool made with 5-imidazolyl-modified uridine.

Another ribozyme carrying 5-pyridyl-uridine residues was isolated using in vitro selection for its ability to form carbon-carbon bonds (Tarasow et al. 1997). The ribozyme promotes a Diels-Alder reaction between an acyclic diene and a maleimide dienophile tethered to the ribozyme and a biotin moiety, respectively (Fig. 8). The reactants were kept apart from the nucleic acid portion of the ribozymes by long-chain polyethyleneglycol linkers. Therefore, a simple template alignment of the reactive groups by Watson-Crick base pairing is not possible. Instead, the selected molecules must form a distinct substrate-binding pocket analogous to those found in the active site of protein enzymes. As with the amide synthases, this enzyme also requires $Cu^{2+}$ to display full catalytic activity. The primary role for divalent metal ions in the action of these ribozymes could be RNA structure stabilization, perhaps through direct coordination with the base modifications. However, the selective dependence on $Cu^{2+}$ also may be due to its Lewis acid properties, which are known to play an important role in catalysis of Diels-Alder reactions in aqueous media (OTTO and ENGBERTS 1995; OTTO et al. 1996).

## 5.2 Modified Oligonucleotides as Components of Reactive Complexes

Some of the most unique RNA and DNA 'enzymes' have been reported by SMITH et al. (1995) and by CHARLTON et al. (1997a). These 'reactive aptamers' accelerate the coupling of a chemically modified oligonucleotide to the active site of neutro-

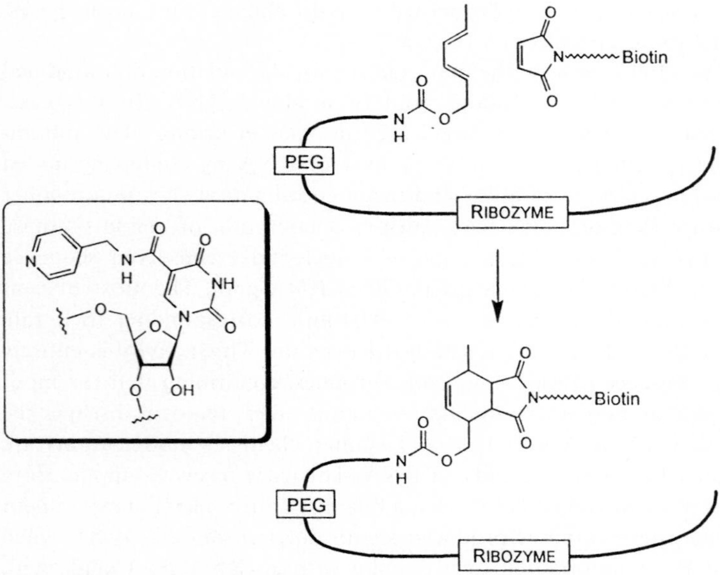


**Fig. 8.** Ribozyme-catalyzed carbon-carbon bond formation via a Diels-Alder reaction mechanism. A random-sequence RNA pool carrying 5-pyridyl-uridine residues (*inset*) is modified with an acyclic diene that is appended to a 5′-terminal polyethyleneglycol linker (PEG). The RNA pool is incubated with a maleimide dienophile conjugated to a biotin moiety. RNAs that promote the coupling reaction are isolated by streptavidin affinity chromatography

phil elastase, thereby serving as extremely potent inhibitors of this protease. In the latter case, a short DNA oligonucleotide modified with a valine phosphonate group was included with a random-sequence DNA pool during in vitro selection for aptamers to neutrophil elastase. Each random-sequence DNA also carried a sequence domain that was complementary to the modified oligonucleotide that allowed complex formation to occur. This 'blended SELEX' protocol yielded an aptamer domain which guided the reactive phosphonate into the active site of human neutrophil elastase, thereby irreversibly blocking the catalytic serine site (Fig. 9). In a subsequent study (CHARLTON et al. 1997b), the additional modification of the splint DNA with fluorescent and radioactive markers allowed diagnostic imaging of inflammation in animal models, thereby providing a practical alternative to antibody-based diagnostic strategies.

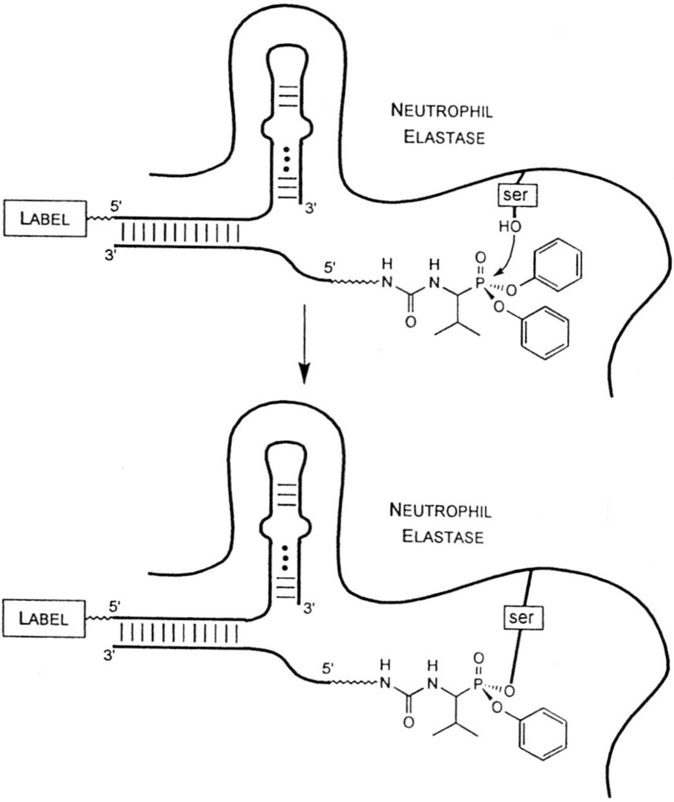


**Fig. 9.** Irreversible inhibition of neutrophil elastase using a reactive DNA aptamer. The hairpin-like DNA aptamer domain guides a separate phosphonate-modified oligonucleotide into the active site of the serine protease, thereby accelerating the transesterification reaction that irreversibly blocks the function of the protein enzyme

# 6 New Strategies for Design and Selection of Nucleic Acid Catalysts

A critical aspect to the continued and accelerated progress with ribozyme and deoxyribozyme engineering is the development of new design strategies. The different selection strategies used to isolate the artificial enzymes discussed in the preceding text all have three features in common. First, these selection schemes demand that each successful enzyme undergo a self-modification event that subsequently is exploited by the system to identify functional molecules. Second, each of the selection protocols is punctuated by the requirement for periodic manipulations such as gel electrophoresis, chromatography, or the preparation of enzymatic modification reactions. More complex selection protocols that include numerous manipulations can become exceedingly time consuming. Third, the process of evolution, either when proceeding naturally or when contained within a test tube, can yield almost any practical mechanism for the molecular 'survival-of-the-fittest' challenges that are presented. Great care must be taken when designing in vitro selection protocols that yield the result of interest while preventing unwanted mechanisms from emerging.

These essentially universal characteristics of existing in vitro selection protocols restrict both the catalytic and kinetic diversity of engineered ribozymes and limit the speed at which the entire selection process can be carried out. For example, the fact that no high-speed multiple-turnover ribozyme has been created to operate on small organic substrates is most likely due to the lack of an effective selection protocol needed to achieve this goal and is not necessarily due to any inherent deficiency of ribozyme function. New design strategies that overcome the limitations of existing protocols would facilitate the continued functional diversification of nucleic acid enzymes. Creating improved in vitro selection protocols that allow total and unlimited control over the design process will not be trivial. However, described below are two recent developments that offer novel or improved methods for ribozyme and deoxyribozyme engineering.

## 6.1 Continuous Evolution

As mentioned above, most in vitro selection protocols follow an iterative stepwise approach, including repeated rounds of selection and amplification requiring periodic manipulations (Williams and Bartel 1996; Breaker 1997a). This process is usually labor intensive and slow, thereby limiting the amount of selection cycles that can be easily achieved to a fairly small number. At least for one class of reactions, this obstacle has been overcome with a 'continuous evolution' strategy (Breaker and Joyce 1994a; Wright and Joyce 1997) that makes use of the self-sustained sequence replication (3SR) reaction (Guatelli et al. 1990; Compton 1991). An RNA ligase ribozyme previously isolated by a more conventional selection protocol (Bartel and Szostak 1993; Ekland et al. 1995) is incubated with a short oligonucleotide that contains a T7 promoter sequence. Only those sequences that successfully acquired this promoter through ligation led to their own reproduction by transcription with T7 RNA polymerase (Fig. 10).

Over the course of an estimated 300 molecular 'generations' of continuous evolution, mutants with improved ligation rates gained a selective advantage in replication and dominated the final population of amplifying RNAs (WRIGHT and JOYCE 1997). The results of this experiment are quite impressive: the isolated ribozyme variants showed an improvement in catalytic rate up to a factor of approximately 14,000. The success of this in vitro evolution experiment, however, greatly depends on the starting pool that is used. A ligase ribozyme that already had robust catalytic activity was required to attain continuous ribozyme-dependent amplification of RNA. A previous attempt starting with a less active group II ribozyme were found to produce selfish RNAs that replicate by a mechanism that bypasses the desired ligation reaction (BREAKER and JOYCE 1994a). In addition, this specific selection strategy will be valuable only with ribozyme reactions that generate an RNA or DNA template that can be used by the polymerases needed for amplification.

## 6.2 Modular Rational Design of Ribozymes

Another improvement in ribozyme engineering strategy has been achieved by fusing preexisting RNA domains to form a chimeric arrangement that makes concerted

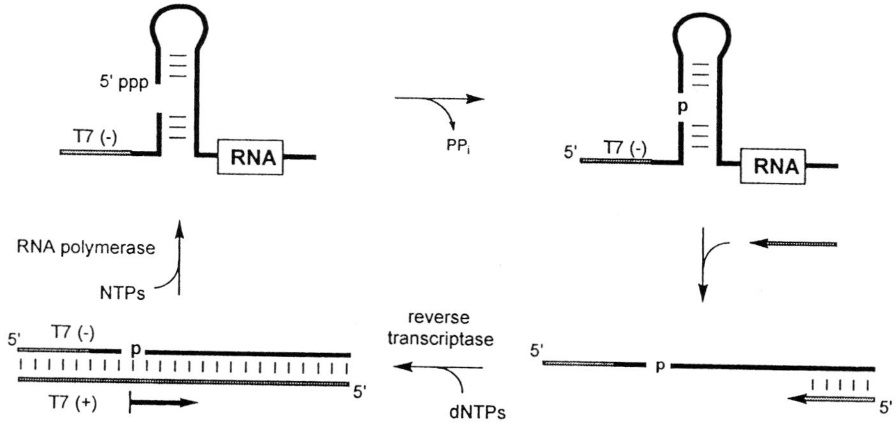


**Fig. 10.** Amplification cycle for the continuous evolution of RNA ligase ribozymes. RNAs that catalyze self-ligation to an oligonucleotide containing the T7 promoter serve as templates for the formation of a cDNA that can be transcribed by T7 RNA polymerase. The newly formed RNA transcripts immediately re-enter this cycle, thereby continuing the selective amplification process without the need for intervention. Ribozymes that acquire mutations that improve ligase function or that gain an advantage at other potential rate limiting steps of the amplification process will dominate the population of RNAs. RNA and DNA is represented by solid and shaded bars, respectively. (Adapted from BREAKER and JOYCE 1994; WRIGHT and JOYCE 1997)

use of their independent functions. Initial success using this approach was reported by Lorsch and Szostak (1994), who created a unique RNA pool by joining a mutagenized ATP-binding aptamer RNA (Sassanfar and Szostak 1993) to random-sequence domains. This pool was used to isolate a number of novel 'RNA kinase' ribozymes, some that apparently make use of the preformed ATP-binding domain to recognize and position the ATP substrate. This biased-pool approach should be widely applicable to generate increasingly complex ribozymes from more simple RNA domains.

This same ATP-binding RNA domain was used to construct a modified hammerhead ribozyme that acts as a true allosteric ribozyme (Tang and Breaker 1997a). This conjoined aptamer/ribozyme complex (Fig. 11a) is arranged such that when ATP is present, the resulting conformational change induced by ligand binding causes a steric clash between aptamer and ribozyme domains (Tang and Breaker 1998). Specifically, in the ATP-bound state the aptamer domain adopts a rigid conformation in which stem IV residing within the aptamer sterically interferes with stem I of the ribozyme, thereby preventing the hammerhead domain from adopting its active structure (Fig. 11b).

This and other related strategies for the modular rational design or the combined modular/combinatorial design of ribozymes have produced a number of allosteric ribozymes that display activation or inhibition ratios of more than 1,000-fold (Soukup and Breaker, in preparation). This magnitude of rate control, coupled with precise ligand-mediated specificity for ribozyme regulation, conceivably could be exploited for a variety of applications. For example, during preparative in vitro transcription of the ATP-sensitive ribozyme depicted in Fig. 11a, its catalytic function remains suppressed due to the presence of ATP in the transcription buffer. Upon gel purification and removal of ATP, ribozyme activity is restored and self cleavage occurs. This new kinetic feature was exploited in a successful effort to conduct in vitro selection on the parental hammerhead ribozyme domain (Tang and Breaker 1997b). The core of the hammerhead ribozyme was randomized and subjected to several rounds of selection for self cleavage without the loss of the most active ribozymes during preparation by in vitro transcription. The sequences of the predominant ribozymes that were selected in this experiment precisely matched the consensus sequence of the hammerhead ribozyme, thereby suggesting that nature has already evolved this motif for optimal catalytic performance. Besides the advantages of such ribozyme constructs for in vitro selection experiments, they might possess a significant application potential in medicine or biotechnology as rate-controlled enzymes.

## 7 Outlook

As revealed in this review, the last several years has provided us with a vast number of stimulating new developments in the field of nucleic acid enzymes. For

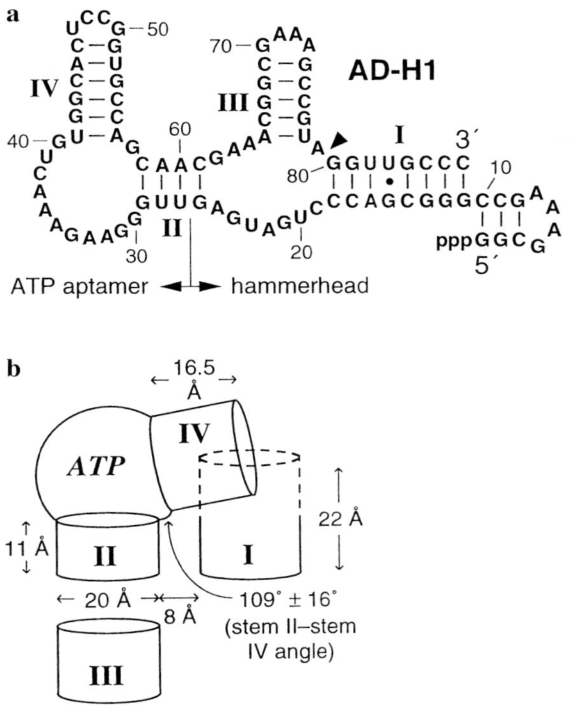


**Fig. 11a,b.** The introduction of allosteric control by modular rational design with aptamers and ribozymes. The conjoined aptamer/ribozyme construct (**a**) and related structures act as ATP-dependent allosteric ribozymes. The mechanism (**b**) involves the mutually exclusive formation of aptamer and ribozyme domains. The relative 3-D orientations of stems I-IV are depicted as *cylinders*. ATP binding to the aptamer domain (spanning stems II and IV) induces a conformational change that results in a clash between IV and stem I of the ribozyme domain, thereby disrupting catalytic function. (Adapted from Tang and Breaker 1997b, 1998)

the near future, we predict the discovery of a seemingly endless stream of new RNA and DNA enzymes that will further expand the repertoire of exciting catalytic capacities. Towards this end, new and more sophisticated selection protocols need to be developed to keep pace with the increasing demands for faster catalytic rates, non-standard reaction conditions and improved or diversified catalytic functions. Already, ribozyme engineers are no longer limited to the catalytic repertoire of standard nucleic acids. The introduction of chemical modifications or the recruitment of small organic molecules as cofactors will lead to dramatic enhancements in the catalytic performance of their parent RNA and DNA molecules. The new generations of nucleic acid enzymes will also show activities that can be tightly regulated by the proper choice of reaction conditions or by small effector molecules, thus opening new dimensions for possible application of these molecules in diagnostics, medicine, biotechnology, or as fundamental tools for organic synthesis.

*Acknowledgements.* This work was supported by a Beckman Young Investigator Award and a Hellman Family Fellowship to RRB. MK also was supported by a postdoctoral fellowship from the German Academic Exchange Service (DAAD).

# References

Bartel DP, Szostak JW (1993) Isolation of new ribozymes from a large pool of random sequences. Science 261:1411–1418

Beaudry AA, Joyce GF (1992) Direct evolution of an RNA enzyme. Science 257:635–641

Benner SA, Ellington AD, Tauer A (1989) Modern metabolism as a palimpsest of the RNA world. Proc Natl Acad Sci USA 86:7054–7058

Breaker RR (1997a) In vitro selection of catalytic polynucleotides. Chem Rev 97:371–390

Breaker RR (1997b) DNA aptamers and DNA enzymes. Curr Opin Chem Biol 1:26–31

Breaker RR (1997c) DNA enzymes. Nature Biotechnol 15:427–431

Breaker RR, Joyce GF (1994a) Emergence of a replicating species from an in vitro RNA evolution reaction. Proc Natl Acad Sci USA 91:6093–6097

Breaker RR, Joyce GF (1994b) A DNA enzyme that cleaves RNA. Chem Biol 1:223–229

Breaker RR, Joyce GF (1995) A DNA enzyme with $Mg^{2+}$-dependent RNA phosphoesterase activity. Chem Biol 2:655–660

Burmeister J, von Kiedrowski G, Ellington AD (1997) Cofactor-assisted self-cleavage in DNA libraries with a 3′-5′-phosphoramidate bond. Angew Chem Int Ed 36:1321–1324

Buzayan JM, Gerlach WL, Bruening G (1986) Non-enzymatic cleavage and ligation of RNAs complementary to a plant virus satellite RNA. Nature 323:349–353

Carmi N, Balkhi SR, Breaker RR (1998) Cleaving DNA with DNA. Proc Natl Acad Sci USA 95:223–237

Carmi N, Shultz LA, Breaker RR (1996) In vitro selection of self-cleaving DNAs. Chem Biol 3:1039–1046

Chapman KB, Szostak JW (1995) Isolation of a ribozyme with 5′-5′ ligase activity. Chem Biol 2:325–333

Charlton J, Kirschenheuter GP, Smith D (1997a) Highly potent irreversible inhibitors of neutrophil elastase generated by selection from randomized DNA-valine phosphonate library. Biochemistry 36:3018–3026

Charlton J, Sennello J, Smith D (1997b) In vivo imaging of inflammation using an aptamer inhibitor of human neutrophil elastase. Chem Biol 4:809–816

Chowrira BM, Berzal-Herranz A, Burke JM (1993) Ionic requirements for RNA binding, cleavage, and ligation by the hairpin ribozyme. Biochemistry 32:1088–1095

Cochran AG, Schultz PG (1990) Antibody-catalyzed porphyrin metallation. Science 249:781–783.

Compton J (1991) Nature 350:91–92

Conn MM, Prudent JR, Schultz PG (1996) Porphyrin metalation catalyzed by a small RNA molecule. J Am Chem Soc 118:7012–7013

Couenoud B, Szostak JW (1995) A DNA metalloenzyme with DNA ligase activity. Nature 375:611–614

Dahm SC, Uhlenbeck OC (1991) Role of divalent metal ions in the hammerhead RNA cleavage reaction. Biochemistry 30:9464–9469

Dai X, De Mesmaeker A, Joyce GF (1995) Cleavage of an amide bond by a ribozyme. Science 267: 237–240

Dai X, De Mesmaeker A, Joyce GF (1996) Amide cleavage by a ribozyme:correction. Science 272:18–19

Eaton BE (1997) The joys of in vitro selection:chemically dressing oligonucleotides to satiate protein targets. Curr Opin Chem Biol 1:10–16

Ekland EH, Bartel DP (1996) RNA-catalysed RNA polymerization using nucleoside triphosphates. Nature 382:373–376

Ekland EH, Szostak JW, Bartel DP (1995) Structurally complex and highly active RNA ligases derived from random RNA sequences. Science 269:364–370

Faulhammer D, Famulok M (1997) Characterization and divalent metal-ion dependence of in vitro selected deoxyribozymes which cleave DNA/RNA chimeric oligonucleotides. J Mol Biol 269:188–202

Frauendorf C, Jaschke A (1998) Catalysis of organic reactions by RNA. Angew Chem Int Ed 37: 1378–1381

Geyer CR, Sen D (1997) Evidence for the metal-cofactor independence of an RNA phosphodiester-cleaving DNA enzyme. Chem Biol 4:579–593

Gilbert W (1986) The RNA world. Nature 319:618

Guatelli JC, Whitefield KM, Kwoh DY, Barringer KJ, Richman DD, Gingeras TR (1990) Isothermal, in vitro amplification of nucleic acids by a multienzyme reaction modeled after retroviral replication. Proc Natl Acad Sci USA 87:1874–1878

Guerrier-Takada C, Gardiner K, Marsh T, Pace N, Altman S (1983) The RNA moiety of ribonuclease P is the catalytic subunit of the enzyme. Cell 35:849–857

Hager AJ, Szostak JW (1997) Isolation of novel ribozymes that ligate AMP-activated RNA substrates. Chem Biol 4:607–617

Hager JA, Pollard JD, Szostak JW (1996) Ribozymes: aiming at RNA replication and protein synthesis. Chem Biol 3:717–725

Hampel A, Cowan JA (1997) A unique mechanism for RNA catalysis: the role of metal cofactors in hairpin ribozyme cleavage. Chem Biol 4:513–517

Hirao I, Ellington AD (1995) Recreating the RNA world. Curr Biol 5:1017–1022

Huang F, Yarus M (1997a) 5′-RNA self-capping from guanosine diphosphate. Biochemistry 36: 6557–6563

Huang F, Yarus M (1997b) Versatile 5′ phosphoryl coupling of small and large molecules to an RNA. Proc Natl Acad Sci USA 94:8965–8969

Huang F, Yarus M (1997c) A calcium-metalloribozyme with autodecapping and pyrophosphatase activities. Biochemistry 36:14107–14119

Illangasekare M, Sanchez G, Nickles T, Yarus M (1995) Aminoacyl-RNA synthesis catalyzed by an RNA. Science 267:643–647

Jayasena VK, Gold L (1997) In vitro selection of self-cleaving RNAs with a low pH optimum. Proc Natl Acad Sci USA 94:10612–10617

Jenne A, Famulok M (1998) A novel ribozyme with ester transferase activity. Chem Biol 5:23–34

Joyce GF (1992) Selective amplification techniques for optimization of ribozyme function. In: Antisense RNA and DNA, TR Cech ed., Wiley-Liss, New York, pp 353–372

Kruger K, Grabowski PJ, Zaug AJ, Sands J, Gottschling DE, Cech TR (1982) Self-splicing RNA: autoexcision and autocyclization of the ribosomal RNA intervening sequence of Tetrahymena. Cell 31:147–157

Kuimelis RG, McLaughlin RW (1995) Hammerhead ribozyme mediated cleavage of a substrate analogue containing an internucleotidic bridging 5′-phosphorothioate: implications for the cleavage mechanism and the catalytic role of the metal cofactor. J Am Chem Soc 117:11019–11020

Letsinger RL, Wu T (1995) Use of a stilbenecarboxamide bridge in stabilizing, monitoring, and photochemically altering folded conformations of oligonucleotides. J Am Chem Soc 117:7323–7328

Li Y, Sen D (1996) A catalytic DNA for porphyrin metallation. Nat Struct Biol 3:743–747

Li Y, Sen D (1998) The modus operandi of a DNA enzyme: enhancement of substrate basicity. Chem Biol 5:1–12

Lin Y, Qiu Q, Gill SC, Jayasena SD (1994) Modified RNA sequence pools for in vitro selection. Nucleic Acids Res 22:5229–5234

Lohse PA, Szostak JW (1996), Ribozyme-catalysed amino-acid transfer reactions. Nature 381:442–444

Lorsch JR, Szostak JW (1994) In vitro evolution of new ribozymes with polynucleotide kinase activity. Nature 371:31–36

Macaya RF, Waldron JA, Beutel BA, Gao H, Joesten ME, Yang M, Patel R, Bertelsen AH, Cook AF (1995) Structural and functional characterization of potent antithrombic oligonucleotides possessing both quadruplex and duplex motifs. Biochemistry 34:4478–4492

Nelson JS, Giver L, Ellington AD, Letsinger RL (1996) Incorporation of a non-nucleotide bridge into hairpin oligonucleotides capable of high-affinity binding to the REV protein of HIV-1. Biochemistry 35:5339–5344

Nesbitt S, Hegg LA, Fedor MJ (1997) An unusual pH-independent and metal-ion-independent mechanism for hairpin ribozyme catalysis. Chem Biol 4:619–630

Noller HF, Hoffarth V, Zimniak L (1992) Unusual resistance of peptidyl transferase to protein extraction procedures. Science 256:1416–1419

Osborne SE, Matsumura I, Ellington AD (1997) Aptamers as therapeutic and diagnostic reagents: problems and prospects. Curr Opin Chem Biol 1:5–9

Osborne SE, Volker S, Stevens SY, Glick KJ (1996) Design, synthesis and analysis of disulfide cross-linked DNA duplexes. J Am Chem Soc 118:11993–12003

Otto S, Bertoncin F, Engberts JBFN (1996) Lewis acid catalysis of Diels-Alder reactions in water. J Am Chem Soc 118:7702–7707
Otto S, Engberts JBFN (1995) Lewis acid catalysis of Diels-Alder reactions in water. Tetrahedron Lett. 36:2645–2648
Pan T, Uhlenbeck OC (1992) In vitro selection of RNAs that undergo autolytic cleavage with $Pb^{2+}$. Biochemistry 31:3887–3895
Peebles CL, Perlman PS, Mecklenburg KL, Petrillio ML, Tabor JH, Jarell KA, Cheng H-L (1986) A self-splicing RNA excises an intron lariat. Cell 44:213–223
Prody GA, Bakos JT, Buzayan JM, Schneider IR, Bruening G (1986) Autolytic processing of dimeric plant virus satellite RNA. Science 231:1577–1580
Prudent JR, Uno T, Schultz PG (1994) Expanding the scope of RNA catalysis. Science 264:1924–1927
Pyle AM (1993) Ribozymes: a distinct class of metalloenzymes. Science 261:709–714.
Roth A, Breaker RR (1998) An amino acid as cofactor for a catalytic polynucleotide. Proc Natl Acad Sci USA 95:6027–6031
Santoro SW, Joyce GF (1997) A general purpose RNA-cleaving DNA enzyme. Proc Natl Acad Sci USA 94:4262–4266
Sassanfar M, Szostak JW (1993) An RNA motif that binds ATP. Nature 364:550–553
Saville BJ, Collins RA (1990) A site-specific self cleavage reaction performed by a novel RNA in Neurospora Mitochondria. Cell 61:685–696
Sharmeen L, Kuo MYP, Dinter-Gottlieb G, Taylor J (1988) Antigenomic RNA of human hepatitis delta virus can undergo self-cleavage. J Virol 62:2674–2679
Smith D, Kirschenheuter GP, Charlton J, Guidot DM, Repine JE (1995) In vitro selection of RNA-based irreversible inhibitors of human neutrophil elastase. Chem Biol 2:741–750
Suga H, Lohse PA, Szostak JW (1998) Structural and kinetic characterization of an acyl transferase ribozyme. J Am Chem Soc 120:1151–1156
Symons RH (1992) Small catalytic RNAs. Annu Rev Biochem 61:641–671
Tang J, Breaker RR (1997a) Rational design of allosteric ribozymes. Chem Biol 4:453–459
Tang J, Breaker RR (1997b) Examination of the catalytic fitness of the hammerhead ribozyme by in vitro selection. RNA 3:914–925
Tang J, Breaker RR (1998) Mechanism for allosteric inhibition of an ATP-sensitive ribozyme. Nucleic Acids Res (in press)
Tarasow TM, Tarasow SL, Eaton BE (1997) RNA-catalysed carbon-carbon bond formation. Nature 389:54–57
Usman N, Beigelman L, McSwiggen JA (1996) Hammerhead ribozyme engineering. Curr Opin Struct Biol 1:627–533
Walsh C (1979) In: Enzymatic Reaction Mechanisms. WH Freeman, New York, pp.199–207
Wecker M, Smith D, Gold L (1996) In vitro selection of a novel catalytic RNA: characterization of a sulfur alkylation reaction and interaction with a small peptide. RNA 2:982–994
Wiegand TW, Janssen RC, Eaton BE (1997) Selection of RNA amide synthases. Chem Biol 4:675–683
Williams KP, Bartel (1996) In vitro selection of Catalytic RNA. Nucleic Acids Molec Biol 10:367–381.
Williams KP, Ciafre S, Tocchini-Valentini GP (1995) Selection of novel $Mg^{2+}$-dependent self-cleaving ribozymes. EMBO J 14:4551–4557
Wilson C, Szostak JW (1995) In vitro evolution of a self-alkylating ribozyme. Nature 374:777–782
Wright MC, Joyce GF (1997) Continuous in vitro evolution of catalytic function. Science 276:614–617
Yarus M (1993) How many catalytic RNAs? Ions and the Cheshire cat conjecture. FASEB J 7:31–39
Zhang B, Cech TR (1997) Peptide bond formation by in vitro selected ribozymes. Nature 390:96–100
Zimmerly S, Guo H, Eskes R, Yang J, Perlman PS, Lambowitz AM (1995) A group II intron RNA is a catalytic component of a DNA endonuclease involved in intron mobility. Cell 83:529–538

# Dynamic Combinatorial Chemistry: Evolutionary Formation and Screening of Molecular Libraries

A.V. Eliseev[1] and J.M. Lehn[2]



## 1 Introduction

The recent rapid development of new approaches to in vitro artificial evolution of biopolymers has occurred simultaneously with the explosive growth of combinatorial chemistry of small synthetic molecules. This trend can hardly be a mere coincidence; rather, it reflects a growing interest of the chemical and biochemical communities in exploring various aspects of molecular diversity generation and screening.

Since the early consideration of peptide libraries (Furka et al. 1988), organic combinatorial chemistry has been forcefully developed in the pharmaceutical industry, aimed at the rapid synthesis and screening of arrays of low-molecular weight compounds in the search for new drug candidates (Balkenhohl et al. 1996). Most often, these arrays are created as libraries of individual compounds rather than their mixtures (pools). While the mixture formation is much more straightforward and considerably less time-consuming than library synthesis, the difficulty of screening and identifying the mixture components impedes practical applications of the pool-based techniques. In contrast, nature commonly manipulates highly populated pools of compounds, such as proteins and antibodies, using evolutionary refined pathways for their selection and amplification.

[1] Department of Medicinal Chemistry, School of Pharmacy, State University of New York at Buffalo, Buffalo, NY 14260, USA
[2] Laboratoire de Chimie Supramoléculaire, ISIS-ULP-CNRS ESA-7006, 4 rue Blaise Pascal, 67000 Strasbourg, France

This review highlights several recent efforts that explore novel types of non-biological combinatorial pools capable of changing their composition by recombination of a set of basic components into a diversity of constituents, as directed by the presence of a molecular target. Such targeted organization is achieved by mimicking some principles (but not necessarily mechanisms) of natural evolution in the synthetic, small molecule libraries. The emergence of this approach is also linked to the development of supramolecular chemistry, involving the implementation of noncovalent interactions for effecting molecular recognition, self-assembly and self-organization processes (LEHN 1995).

## 2 The Concept

The basic idea of the self-selecting combinatorial pools is depicted in Fig. 1. In contrast to the "traditional" static combinatorial approach, the key design feature is that the library constituents exist in a thermodynamic equilibrium that dynamically maintains chemical diversity of the mixture. If the target compound (template) is then added to the equilibrating pool, the equilibrium in the system will be shifted toward a higher fraction of the constituent(s) ($A_i$) that form(s) stronger complex(es) with the target ($A_iT$). The subsequent quenching of the equilibration process and dissociation of the target-constituent complexes results in the isolation of the enriched library in which the amount of effective binders is amplified at the expense of the noneffective ones. The above transformation of the equilibrated mixture bears resemblance to a Darwinian evolution process in that the fittest library members, i.e. the best binders, are produced in greater amounts while the population of weaker binders decreases in the targeted self-selection process. As will be shown below, this evolutionary approach can be implemented in a variety of chemical libraries and targets. Such equilibrating mixtures may be referred to as "dynamic libraries" and, in effect, as "virtual libraries" (HUC and LEHN 1997) to indicate that the dynamic process makes all combinations potentially accessible even if they may well not be present in significant amounts in absence of the target. The "virtuality" notion thus extends the simply "dynamic" character beyond the mixture of equilibrating constituents of the library.

The remarkable simplicity and generality of the above concept did not go unnoticed over the years since the underlying Le Chatelier principle has been known. Indeed, the concept of the selective equilibrium shift has been applied, for instance, to isomeric mixtures of steroids (WOODWARD et al. 1952), porphyrins (ELLIOTT 1980; LINDSEY 1980) and natural ionophores (STILL 1987) to shift an equilibrium to preferential formation of the desired isomers. The thermodynamic template effect has also been used in numerous examples of self-assembling systems (LEHN 1995; FUJITA 1995; HAMILTON 1997) to drive noncovalent association of molecular recognition elements to well-defined supramolecular architectures. A reversible thermodynamically controlled pre-assembly of complementary

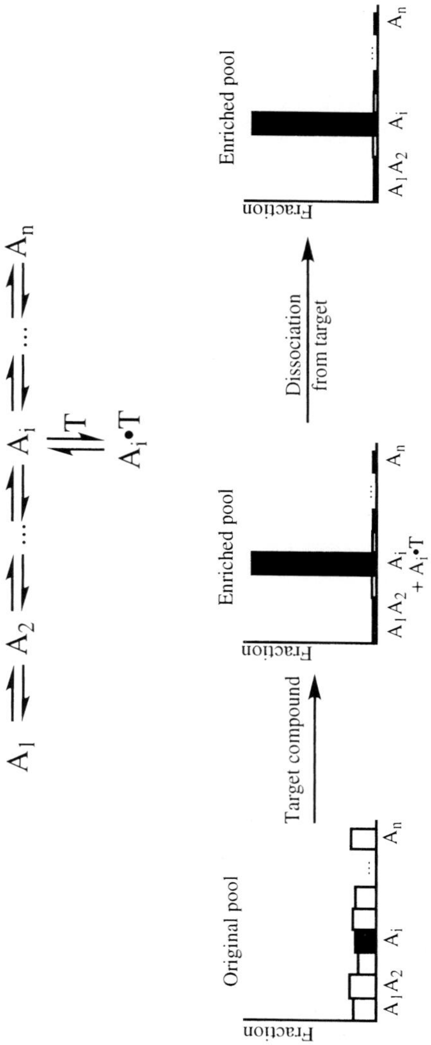


**Fig. 1.** Target-induced self-organization of a dynamic library

nucleotides has been shown to promote selective chemical coupling of DNA fragments (GOODWIN and LYNN 1992; ZHAN and LYNN 1997). Molecular imprinting of small molecules into three-dimensional polymeric structures leading to the formation of new media for molecular recognition and catalysis (MOSBACH and RAMSTROM 1996) has also been based on certain degree of selective preassociation of components.

It was not until very recently, however, that the concept of the target-driven organization of molecular diversity has crystallized with respect to its implementation in diversity generation and its combinatorial utility. The recent contributions discussed below present additional key features with respect to the earlier studies on template-directed equilibration:

1. The dynamic combinatorial mixtures represent *diverse* populations of compounds, in terms of both structural characteristics and the numbers of components. While most of the initial experiments have been performed with relatively small pools, ranging from just a few to several dozens of compounds, the method is potentially applicable to much larger libraries that cover a significant amount of diversity space.
2. The constituents of the dynamic libraries can either be isolated and characterized as individual compounds already in absence of the target, or only be expressed on binding to the target. In either case, they can be subsequently synthesized on a larger scale.
3. The dynamic libraries are designed with a specific molecular target in mind. Because the ultimate goal of the dynamic approach is to discover lead compounds that recognize the target in solution, particular attention is also paid to the selection conditions that should be similar to those where the effective binders will be used.

## 3 Specific Examples

The strategies used so far to exploit the concept of dynamic diversity, can be divided into two categories. In one set of approaches, the library components are stable under the selection conditions, and the equilibration reaction (Fig. 1) occurs separately from the selection process.

The viability of this strategy was first demonstrated by using a monoclonal antibody (3E7) as a molecular sink capable of directing reversible synthesis/hydrolysis of a peptide mixture towards the formation of the specific antibody binder (VENTON et al. 1994; SWANN et al. 1996). The dynamic mixture was generated from two initial peptides, YGG and FL, that had been previously shown to form a combination of both coupling and cleavage products in the presence of a proteolytic enzyme, thermolysin. Among the mixture components, identified by HPLC and sequencing, was the pentapeptide YGGFL, known to bind 3E7 with an affinity constant of 7.1nM. The antibody solution used for selection was placed in a dialysis bag and immersed in the dynamic mixture (Fig. 2). The equilibrating peptides could freely diffuse through the dialysis membrane and bind to the antibody that would thus remove the effective components from the equilibrium. Direct measurement of the antibody effect upon the equilibrium was hampered, because the antibody amount used for selection was low enough to be saturated by YGGFL formed even in the unbiased equilibrium. For this reason, the authors estimated the relative amount of the effective binder that migrated into the dialysis bag from competitive displacement of radioactively labeled β-endorphin (specific antigen of 3E7) from its antibody complex. A set of five independent experiments showed that equilibration of the peptide mixture coupled with the antibody had led to an average 12% increase in the displaced endorphin amount as compared to the uncoupled reaction. Thus the authors were able to demonstrate that the presence of a macromolecular sink in the form of the 3E7 antibody was able to amplify the formation of YGGFL or YGGFL-like binding affinity from the protease-based synthetic reaction.

A different, stepwise implementation of the selection/equilibration processes in an automated setup was applied to the evolutionary formation of arginine binders (Eliseev and Nelen 1997). A mixture of three *cis,trans* isomeric dicarboxylates (Fig. 3a) contained one component (*cis,cis* isomer) with a higher affinity to guanidinium derivatives than the other two isomers. All three components, were stable under ambient conditions, but could be interconverted upon irradiation with UV light. The mixture, which initially contained only a minor amount of the *cis,cis* component, was subjected to circulation in the apparatus shown in Fig. 3b. Arginine, used as the target compound, was immobilized in the affinity column (selection chamber). Every pass of the mixture through the immobilized target depleted the solution phase of the *cis,cis* isomer. The *cis,cis* isomer was regenerated upon passing the "mutation" chamber due to photochemical isomerization of the remaining *trans,trans* and *cis,trans* components. Repetition of these cycles led to transformation of the majority of original solution mixture to one highly enriched with the *cis,cis* isomer which accumulated on the immobilized arginine. The *total* amount of *cis,cis* isomer obtained after the experimental cycles was 75% higher

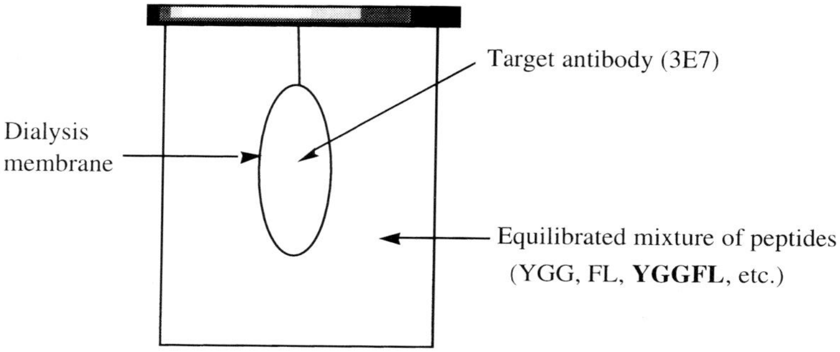


**Fig. 2.** Equilibrium shift in peptide synthesis/hydrolysis in the presence of the target antibody

than that obtained upon photoequilibration of the original mixture in solution. Irradiation of the mixture in the presence of the target *in solution* led to noticeably smaller degree of amplification than the "evolutionary" cycles (ELISEEV and NELEN 1998).

As already mentioned, the diversity generation in the above approaches was performed separately from the selection process. It has been demonstrated that the eventual population of the pools in those cases is determined by the set of thermodynamic constants that characterize equilibria shown in Fig. 1 (ELISEEV and NELEN 1998). An essential feature of these methods is that the pool components exist and can be selected as stable compounds unless equilibrating conditions, such as catalysis, irradiation, and heating, are applied.

An alternative strategy makes use of dynamic combinatorial chemistry (DCC) to gain access to virtual combinatorial libraries (VCL) (HUC and LEHN 1997). It invokes a reversible process of self-assembly of the library components either through noncovalent forces or via a reversible chemical reaction. If given members of a virtual library possess preferential affinity to a specific target, addition of the latter will shift the equilibrium towards expression of the stronger binders. Two

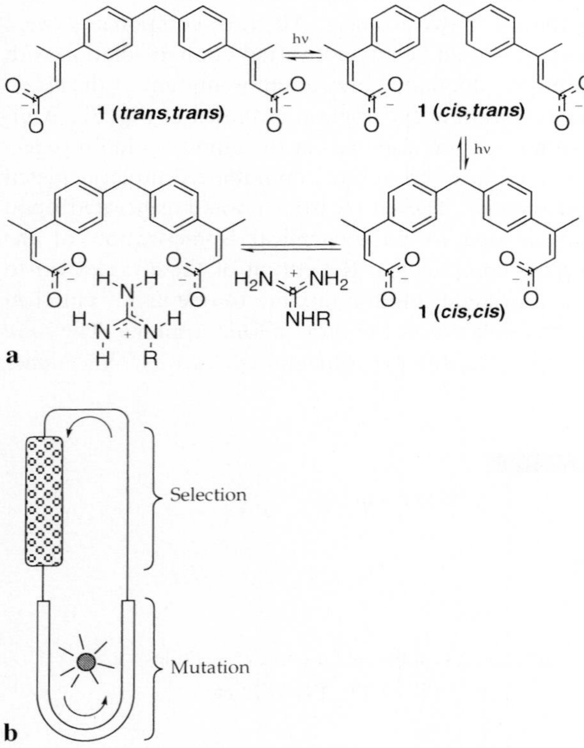


**Fig. 3a,b.** Chemical evolution of arginine receptors from a photoequilibrated mixture. **a** Target-driven chemical equilibrium; **b** experimental setup for continuous "evolution" of effective binders

modes of such target-driven self-assembly, "casting" and "molding" have been proposed, which correspond to assembling substrates and receptors, respectively (Fig. 4).

The DCC/VCL concept was first introduced in the context of the generation of circular helicates, circular double helical metal complexes capable of binding an anionic substrate (HASENKNOPF et al. 1996, 1997). It was shown that the presence of chloride led the equilibrating mixture to converge to the exclusive generation of a pentanuclear chloride complex which was not formed in presence of anions such as sulfate or tetrafluoro borate.

The first bioorganic demonstration of the concept made use of a library of Schiff bases from a set of three aldehydes and four amines (Scheme 1):

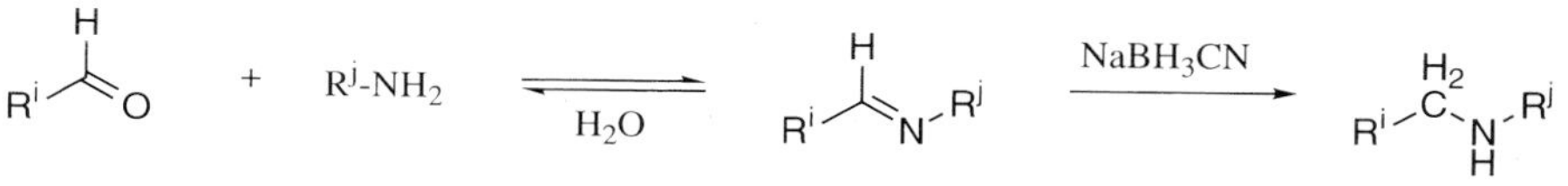


**Scheme 1.** Dynamic library of potential carbonic anhydrase inhibitors

The amine and aldehyde units were designed to present structural features close to those of known efficient inhibitors of carbonic anhydrase II (CA). Although the Schiff bases possess low stability in aqueous solutions, they can be

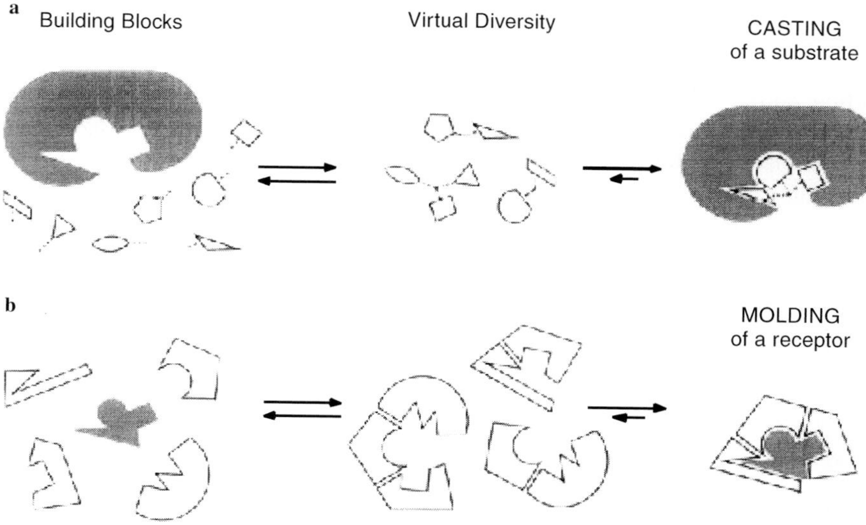


**Fig. 4a,b.** Virtual combinatorial libraries. **a** The casting process: receptor-induced self-assembly of the complementary substrate from a collection of components/fragments. This process amounts to the selection of the optimal substrate from a virtual substrate library. **b** The molding process: substrate-induced self-assembly of the complementary receptor from a collection of structural components/fragments. This process amounts to the selection of an optimal receptor from a virtual receptor library. The diverse potential constituents of the libraries (*center*) are either covalently linked or noncovalently bound reversibly generated species that may or may not exist in significant amount(s) in the free state, in absence of the partner readily reduced to the corresponding amines thereby preserving most of their structural features and allowing one to analyze the library composition. Two sets of experiments, wherein the virtual library was formed in the presence and in the absence of the target enzyme followed by reduction of the components, showed that addition the target markedly changed the abundance of certain compounds in the library. Compound 1 was shown to be amplified by 150% (or up to a factor of about 20 relative to other constituents) in the presence of the enzyme, consistent with the known inhibitory activity of structural analogs of compound 1 toward CA. Importantly, the enzyme ability to bias the equilibrium was diminished in the presence of a known competitive inhibitor of CA, which indicated that it was the enzyme active site that served as a "casting" moiety for the library components.

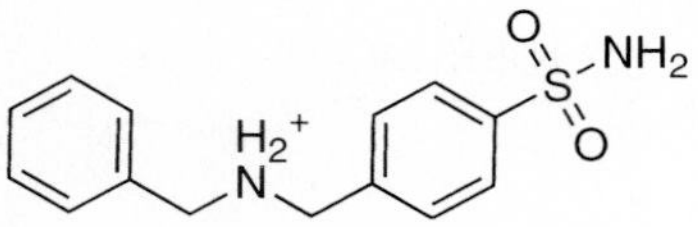

**1**

The method of biasing the composition of a virtual library by casting on a biopolymer structure was further tested on an equilibrated mixture of potential DNA binders composed of zinc complexes of Schiff bases (KLEKOTA et al. 1997). A set of monomeric Schiff base- ligands of general structure 2 was used to form a pool of up to 36 different bidentate Zn(II) complexes (Fig. 5) with potential affinity to the DNA double helix. The target poly(dAT) double helix was immobilized on resin beads and used to select a fraction* of the zinc complexes from the equilibrating pool. A direct analysis of the DNA-bound complexes was not possible, since the conditions used for their dissociation from the target also led to hydrolysis of the complexes. However, the authors designed a set of control experiments, such as determination of the monomer composition in the support-bound fraction in the presence and in the absence of zinc as well as comparison with adsorption on the unmodified support. The results of the control runs allowed them to identify a complex that indeed had a higher affinity to the target duplex as was confirmed by subsequent binding studies in solution.

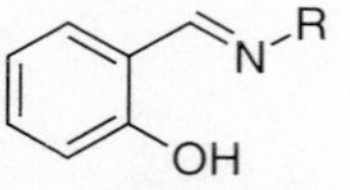

**2**

The approach defined above as "molding" (Fig. 4a) was also used to bias the composition of virtual receptor libraries due to assembly of selected receptor structures around low-molecular targets. In a series of background studies, the Sanders group discovered that mixtures of macrocyclic oligomers, such as oligocholates (Fig. 6), could be formed under thermodynamic control by reversible

---

* The 85:1 complex to base pair ratio used in the experiment seemed to be insufficient to drive the equilibration to completion.

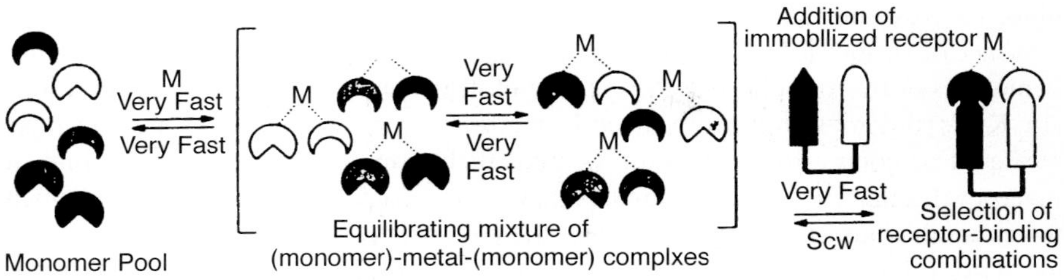


**Fig. 5.** Selection of DNA binders from a dynamic mixture of metal complexes

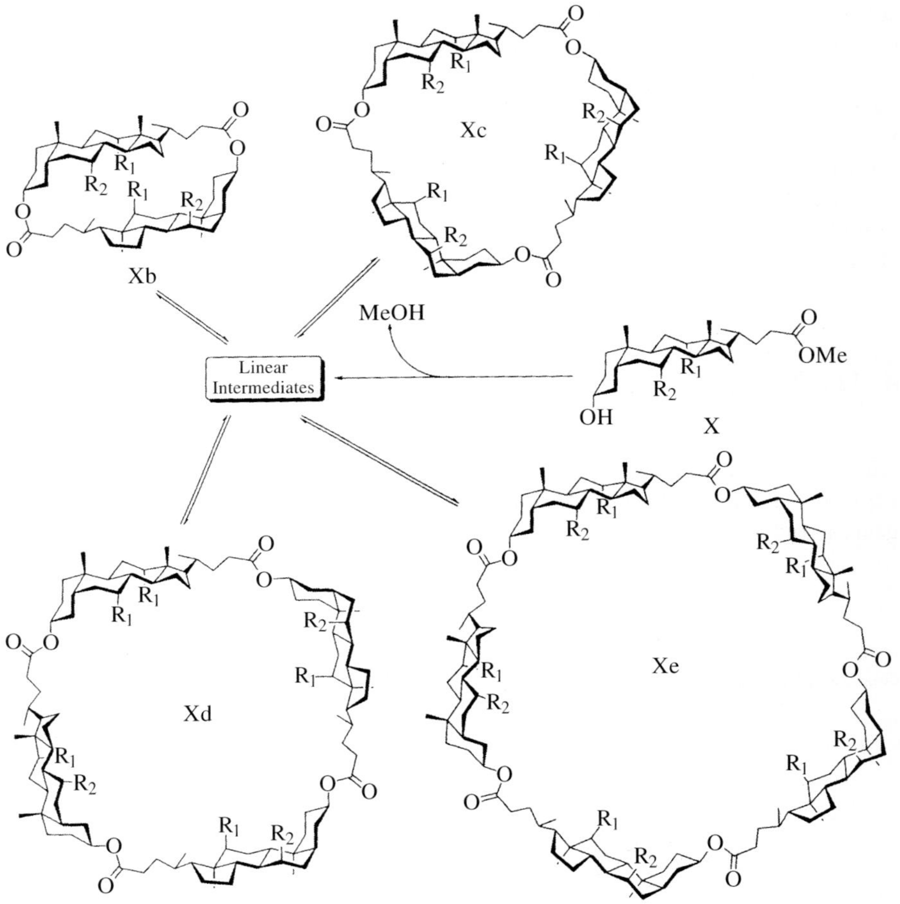


**Fig. 6.** An equilibrating mixture of cyclic steroid oligomers

transesterification reactions (BRADY and SANDERS 1997b). It was proposed that the equilibrium in libraries of macrocycles modified with ionophore polyether side chains could be shifted by the presence of an alkaline metal ion that preferentially binds to one of the macrocycles. The macrocyclization experiments performed in the presence of various ions showed that Na+ favored the formation of cyclic tetra- and pentamers of the ionophore at the expense of its trimers content, whereas other ions (Li, K, Cs) did not significantly affect the mixture composition (Brady and Sanders 1997a). For comparison, the authors studied the ion effect on the equilibrium of macrocycles that did not bear obvious metal ion recognition functionality. A much lower equilibrium shift was observed in these control experiments.

Hioki and Still (1998) used synthetic receptors for peptides, isolated previously from combinatorial libraries, to explore evolutionary formation of the effective peptide binders in equilibrated mixtures of three receptors (Fig. 7). The starting receptor ASSB was formed by linking the ASH and BSH units via a disulfide bond. Applying the thiol-disulfide exchange conditions to ASSB led to its disproportionation to give an equilibrium mixture of 35mol% of ASSB and 65mol% of two products, ASSA, and BSSB. Similar transformation was then performed in the presence of the immobilized peptide Ac(D)Pro(L)Val(D)Val-PS, known to possess particularly high affinity to ASSA. Addition of the peptide increased the yield of ASSA and BSSB to 95mol%, the majority of the ASSA being bound to the peptide-containing resin. A similar amplification of the effective receptor was obtained on disproportionation of ASSC.

## 4 The Niche

Although the early experiments described above represent only initial steps in exploring dynamic diversity, they identify dynamic combinatorial chemistry as a general and promising approach to self-selecting molecular populations of various

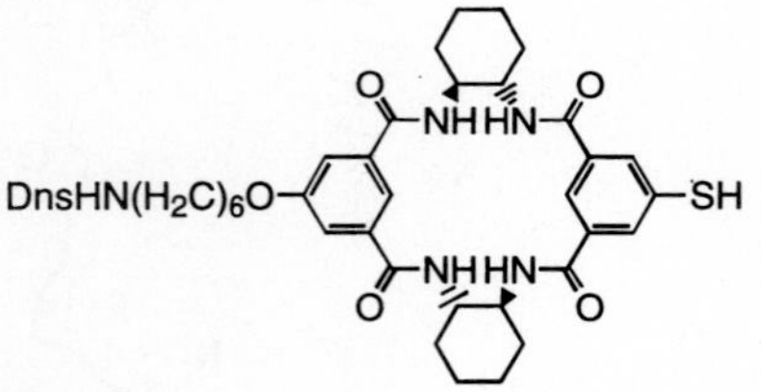


**A**-SH

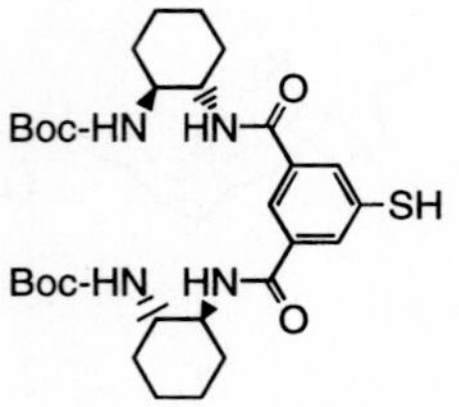

**B**-SH

PhSH **C**-SH

**Fig. 7.** Building blocks for macrocyclic peptide receptors types and functions. The self-selection, or evolution, of the dynamic libraries mimics some of the essential features that are used in manipulations with the biodiversity systems described elsewhere in this volume. It is therefore tempting to compare the dynamic combinatorial approach with existing methods that operate with pools of nucleic acids and peptides. Such a comparison, which also includes attributes of "traditional" combinatorial chemistry, is given in Table 1.

Perhaps one of the most intriguing features of dynamic combinatorial chemistry, although utilized so far only to a limited extent, is its ability to amplify selected components in the libraries. The degree of amplification has certain limits that have been roughly estimated by ELISEEV and NELEN (1998). It is obvious that the extent of equilibrium shift in the dynamic libraries by the target is highly dependent on the selectivity of target binding among the components. The estimate showed that if a hypothetical dynamic library of $n$ equally distributed components contained a single member that possesses $n$-fold higher target affinity than all other members, then the target-induced amplification could lead to the transformation of approximately 50% of the overall library material to the form of effective component. This corresponds to an $n/2$-fold amplification of the best binder. While such an amplification degree is incomparable, for example, with that achieved by PCR, it is applicable to a much larger functional diversity of compounds. More importantly, the amplification in dynamic libraries is selective and increases only the amounts of strong binders, while PCR, unless coupled with an additional selection step (FAMULOK and SZOSTAK 1992), proportionally amplifies all pool

**Table 1.** Comparison of approaches to molecular diversity generation and screening

| Approach | Reference | Library (pool) population | Time frame of lead identification | Functional diversity (number of groups involved) | Relationship between diversity generation and screening |
|---|---|---|---|---|---|
| Static combinatorial chemistry (libraries of individual compounds) | BALKENHOHL et al. 1996 | $10–10^5$ | Months to years | Unlimited | Separate |
| Nucleic acid evolution (SELEX) | FAMULOK and SZOSTAK 1992 | $10^{13}$ | Hours to days | ~4 | Partly combined |
| Mutagenesis-in vivo selection of proteins | KAST and HILVERT 1996 | $10–10^{13}$ | Weeks to months | ~20 | Separate |
| Dynamic combinatorial chemistry (projected) | | $10–10^5$ | Days | Unlimited | Combined |

components. In this context, it may be relevant to coin the word "non-linear amplification" to describe the selective production of effective binders within the dynamic libraries.*

On the other hand, the amplification may be much larger when the dynamic library consists of a variety of oligomeric constituents derived from a single component. This is the case in the quantitative conversion of a helicate VCL into the pentameric chloride binding constituent (HASENKNOPF et al. 1996, 1997).

The dynamic approach shows promise for a practical combinatorial chemist, because it essentially combines diversity generation and screening in a single process. The screening is additionally facilitated because not only the amount of strong binders increases, but also the abundance of the binding background decreases in the dynamic library. The whole process can therefore be considered as a selective informational signal-to-noise increase in a molecular population.

## 5 Quo Vadis?

Dynamic combinatorial chemistry is currently in its early stage of development. Notably, all of the above experiments used different dynamic library chemistry as well as different targets. This suggests that only scattered islands of application for this method have been pinpointed on a potentially rich map. Significant momentum has been gained over the last 2 years in background studies, and considerable interest in further development of the dynamic approach has been generated. A variety of developments may be envisaged, be they organic or inorganic, of chemical, biological or medicinal character. A few possible directions for future growth of this new area can be envisioned, based on the fundamental attributes of the method.

The dynamic approach to molecular diversity represents a potential shift of the synthetic chemistry paradigm. The basis of the dynamic libraries, reversible chemical reactions, have often been looked at as a nuisance by a synthetic chemist because of the poor product yields they can generate. On the contrary, dynamic combinatorial chemistry calls for a nontrivial search for smooth and unselective reversible reactions that can provide equal representation and facile interconversion of all of the library components. Of particular interest here would be new or re-invented bond making and breaking transformations, such as various types of condensations, that can be performed in water, the medium of choice for selection with biological targets (POLYAKOV et al. 1999). Promising equilibrating libraries may be also formed from inorganic coordination compounds (HASENKNOPF et al. 1996, 1997; HAMILTON 1997).

It should be pointed out that, in addition to the various examples of equilibrating *molecular* libraries described above, the target-induced self-assembly of

* The term non-linear reflects the selectivity of amplification among different library components rather than increase of the amount of material with the number of cycles.

specific *supramolecular* entities represents another powerful approach. Such supramolecular VCLs may express either inorganic constituents resulting from the reversible combination of ligand components around metal ions (HASENKNOPF et al. 1996, 1997) or organic supramolecular constituents whose components link together through noncovalent interactions such as hydrogen bonds (CALAMA et al. 1998), van der Waals contacts, electrostatic or donor-acceptor effects, etc.

Given the critical role of target-binding selectivity among the dynamic library components, further effort will likely be directed towards designing the libraries with unique binding properties of each component. This may be achieved by covering a broad functional diversity of the component building blocks as well as careful planning of regio- and stereo-chemistry of the library members. Selectivity may be also controlled by well-defined target structures. Biopolymer targets, such as proteins and nucleic acids, are likely to display a high selectivity for minor structural variations among similar library components and should thus lead to higher pressure on the evolution of the dynamic library. In the long run, selection and evolution processes in the dynamic libraries may also be targeted toward catalysis of chemical reactions, rather than binding events, similar to the selection of catalytic nucleic acids (LORSCH and SZOSTAK 1996) or the self-assembly of inorganic catalysts (HILL 1995). Catalysis may be expected to achieve high degrees of selective amplification in the evolving combinatorial pools of catalysts.

The analogy with natural evolution can be extended by further application of the features of genetic algorithms to the modification of dynamic libraries (WEBER et al. 1995; SINGH et al. 1996). While the primary evolutionary process in the equilibrating pools occurs in an automatic fashion, human intervention may advantageously be performed for combining stepwise improvement of the next-generation libraries with self-selection processes under thermodynamic pressure.

## References

Balkenhohl F, von dem Bussche-Hünnefeld C, Lansky A, Zechel C (1996) Combinatorial synthesis of small organic molecules. Angew Chem Int Ed Engl 35:2288–2337

Brady PA, Sanders JKM (1997a) Thermodynamically-controlled cyclization and interconversion of oligocholates: metal ion templated 'living' macrolactonisation. J Chem Soc Perkin Trans 1:3237–3253

Brady PA, Sanders JKM (1997b) Selection approaches to catalytic systems. Chem Soc Rev 26:327–336

Calama MC, Hulst R, Fokkens R, Nibbering NMM, Timmerman P, Reinhoudt DN (1998) Libraries of non-covalent hydrogen-bonded assemblies; combinatorial synthesis of supramolecular systems. Chem Commun 1021–1022

Eliseev AV, Nelen MI (1997) Use of molecular recognition to drive chemical evolution. 1. Controlling the composition of an equilibrated mixture of simple arginine receptors. J Am Chem Soc 119:1147–1148

Eliseev AV, Nelen MI (1998) "Use of molecular recognition to drive chemical evolution. 2. Mechanisms of an automated genetic algorithm implementation. Chem Eur J 5:823–832

Elliott CM (1980) Isolation of minor equilibrium components by continuous partition applied to substituted tetraphenylporphyrin atropisomers. Anal Chem 52:666–668

Famulok M, Szostak JW (1992) In vitro selection of specific ligand-binding nucleic acids. Angew Chem Int Ed Engl 31:979–988

Fujita M, Nagao S, Ogura K (1995) Guest-induced organization of a three-dimensional palladium(II) cagelike complex. A prototype for "induced-fit" molecular recognition. J Am Chem Soc 117: 1649–1650

Furka A, Sebestyén F, Asgedom M, Dibó G (1988) Cornucopia of peptides by synthesis. Abstr. 14th Int. Congr. Biochem., Prague, 5:47; Abstr. 10th Int Symp Med Chem, Budapest, 288

Furka A, sebestyén F, Asgedom M, Dibó G (1991) General method for rapid synthesis of multicomponent peptide mixtures. Int J Pept Protein Res 37:487–493

Goodwin JT, Lynn DG (1992) Template-directed synthesis:use of a reversible reaction. J Am Chem Soc 114:9197–9198

Hasenknopf B, Lehn JM, Kneisel BO, Baum G, Fenske D (1996) Self-assembly of a circular double helicate. Angew Chem Int Ed Engl 35:1838–1840

Hasenknopf B, Lehn JM, Boumediene N, Dupont-Gervais A, Van Dorsselaer A, Kneisel B, Fenske D (1997) Self-assembly of tetra- and hexanuclear circular helicates. J Am Chem Soc 119:10956–10962

Hill CL, Zhang X (1995) A "smart" catalyst that self-assembles under turnover conditions. Nature 373:324–326

Hioki H, Still WC (1998) Chemical evolution: a model system that selects and amplifies a receptor for the tripeptide (D)Pro(L)Val(D)Val. J Org Chem 63:904–905

Huc I, Lehn JM (1997) Virtual combinatorial libraries: dynamic generation of molecular and supramolecular diversity by self-assembly. Proc Natl Acad Sci USA 94:2106–2110; 94:8272

Kast P, Hilvert D (1996) Genetic selection strategies for generating and characterizing catalysts. Pure Appl Chem 68:2017–2024

Klekota B, Hammond MH, Miller BL (1997) Generation of novel DNA-binding compounds by selection and amplification from self-assembled combinatorial libraries. Tetrahedron Lett 38:8639–8642

Lehn JM (1995) Supramolecular chemistry, concepts and perspectives. VCH, Weinheim New York

Lindsey J (1980) Increased yield of a desired isomer by equilibria displacement on binding to silica gel, applied to *meso*-tetrakis(*o*-aminophenyl)porphyrin. J Org Chem 45:5215

Linton B, Hamilton AD (1997) Formation of artificial receptors by metal-templated self-assembly. Chem Rev 97:1669–1680

Lorsch JR, Szostak JW (1996) Chance and necessity in the selection of nucleic acid catalysts. Acc Chem Res 29:103–110

Mosbach K, Ramstrom O (1996) The emerging technique of molecular imprinting and its future impact on biotechnology. Bio-Technology 14:163–170

Polyakov VA, Nelen MI, Nazarpack-Kandlousy N, Ryabov AD, Eliseev AV (1999) Imine exchange in O-alkyl and O-aryl oximes as a base reaction for aqueous 'dynamic' combinatorial libraries. A kinetic and thermodynamic study. J Phys Org Chem (in press)

Rowan SJ, Sanders JKM (1997) Building thermodynamic combinatorial libraries of quinine macrocycles. Chem Comm 1407–1408

Singh J, Ator MA, Jaeger EP, Allen MP, Whipple DA, Soloweij JE, Chowdhari S, Treasurywala AM (1996) Application of genetic algorithms to combinatorial synthesis: a computational approach to lead identification and lead optimization. J Am Chem Soc 1996, 118:1669–1676

Still WC (1996) Discovery of sequence-specific peptide binding by synthetic receptors using encoded combinatorial libraries. Acc Chem Res 29:155–163

Still WC, Hauck P, Kempf D (1987) Stereochemical studies of lasalocid epimers. Ion-driven epimerizations. Tetrahedron Lett 28:2817–2820

Swann PG, Casanova RA, Desai A, Frauenhoff MM, Urbancic M, Slomczynska U, Hopfinger AJ, Le Breton GC, Venton DL (1996) Nonspecific protease-catalyzed hydrolysis/synthesis of a mixture of peptides: product diversity and ligand amplification by a molecular trap. Biopolymers 40:617–625

Terrett N (1996) Combinatorial chemistry. Drug Discovery Today 1:450

Venton DL, Hopfinger AJ, Le Breton G (1994), Method for generation and screening of useful peptides. US Pat. 5,366,862

Weber L, Wallbaum S, Broger C, Gubernator K (1995) Optimization of the biological activity of combinatorial compound libraries by a genetic algorithm. Angew Chem Int Ed Engl 34:2280–2282.

Woodward RB, Sondheimer F, Taub D, Heusler K, McLamore WM (1952) The total synthesis of steroids. J Am Chem Soc 74:4223–4251

Zhan ZYJ, Lynn DG (1997) Chemical amplification through template-directed synthesis. J Am Chem Soc 119:12420–12421

# Evolutionary Biotechnology – Reflections and Perspectives

U. KETTLING, A. KOLTERMANN, and M. EIGEN



## 1 Introduction

With the designing of biomolecules, a new era in biotechnology has been initiated. One of the most promising strategies for successfully designing complex biomolecular functions is to exploit nature's principles of Darwinian evolution, i.e. variation and selection. The application of these principles to directed evolution of molecules is the underlying concept of evolutionary biotechnology. Important prerequisites are a comprehensive understanding of the mode of molecular evolution as well as the ability to apply its principles to experimental systems in order to create and optimize molecular functions with scientific or economic value.

The invention of techniques for in vitro recombination of genetic information by Stanley Cohen, Herbert Boyer and coworkers, in the early 1970s, are milestones of modern biosciences (COHEN et al. 1973). These developments broke the ground for genetic engineering, thus revolutionizing modern biotechnology. Biomolecules such as enzymes and other proteins, which until then had to be extracted in small amounts from their natural sources, could now be produced by expressing their genes in heterologous hosts. During the following two decades, intense efforts have led to the discovery of new gene products, to the development of efficient cloning strategies and optimal expression cell lines, and to the design of effective fermentation and down-stream processes for large-scale biotechnological production.

Max Planck Institute for Biophysical Chemistry, Dept. Biochemical Kinetics, Am Fassberg, D-37077 Göttingen, Germany

Today, this process of cloning and expression of genes coding for natural proteins with economic value has developed to a high level. A large variety of natural peptides and proteins of pharmaceutical relevance has already been produced in recombinant form with a quality sufficient to be applied as therapeutics. Likewise, recombinant enzymes that are important for diagnostic purposes or as catalysts for industrial processes are available in large quantities and at low cost. In the context of this chapter, we call this biotechnology conservative, because, in the sense of biology, it is conservative as far as products and methods of production are concerned.

During the last several years, however, an increasing number of researchers have recognized that nature provides a large but finite source of potentially useful molecules. Correspondingly, the desire to design molecules for specific purposes emerged. Such designed molecules should be improvements on their natural counterparts rather than just mimic them. The recent development may thus be referred to as "biotech's second generation" (Gibbons 1992). Biomolecular design has indeed redefined the meaning of modern biotechnology. And it has motivated a whole range of new disciplines in the biosciences giving rise to different design strategies: bioinformatics aiming at a rational design using structural information and theoretical calculations, combinatorial chemistry and biochemistry for design on a trial-and-error basis, and evolutionary biotechnology aiming at design as nature does it, i.e. by evolutionary means.

## 2 Designing Biomolecules

Basically, three important questions arise when intending to design biomolecules: (1) Which particular molecular features are of interest? (2) Which class of biomolecule is appropriate to match these features? (3) Which is the most efficient way to generate them? The last question is the most important and crucial one; we shall approach it first from a rather conceptual point of view, outlining the more generic aspects. Practical aspects will be discussed afterwards.

Design of functional molecules is actually generation of information that enables us to construct this function. In terms of biomolecules, information is usually referred to as the genotype, whereas the function (or a combination of several functions and properties) is called the phenotype. All possible genotypes represent the sequence space (Eigen and Schuster 1977), whereas all phenotypical variants together build the functional space (Schuster 1995). Here, we will focus on nucleic acids and proteins, which are the two most important classes of functional biomolecules. A crucial difference between these two classes in the context of biomolecular design is that only nucleic acids are able to act as both genotype carriers and phenotype carriers. In contrast, the genotype of proteins has to be encoded by nucleic acids to be re-amplifiable. In practice, this implies the employment of expression systems and the necessity for a genotype-phenotype linkage.

Basically, a de novo design has the largest potential to create optimal functions. De novo means that the starting point for a molecular design is a random genotype without any known function. Due to the highest degree of freedom, it provides the maximal chance to find the best variant for a certain function. For small molecules such as ligands or inhibitors, de novo design has already been achieved. However, for larger molecules and more complex functions, we struggle with the complexity problem (see below). Therefore, molecular design is more like molecular engineering; it starts from structures with known functions, which then may be modified successively to come up with one or more new functions.

An ultimate goal would be to generate functional information by a computer, followed by chemical synthesis of the molecule using an automated synthesizer. However, the computer program has to solve two rather difficult problems: (1) It has to calculate which three-dimensional structure is responsible for a certain molecular function; and (2) which ensemble of atoms and bonds between them folds into this particular three-dimensional structure under given conditions. The latter problem concerns the connection between sequence space and structure space, whereas the first deals with the connection of structure space and functional space. Unfortunately, we are still far away from such an entirely computer-aided molecular design, despite the fast development in computer technology and an increasing knowledge of structure-to-function relationships. Moreover, confronted, e.g. with the tremendous complexity of dynamic interactions between the atoms of a single protein molecule, we may doubt whether we will be able at all to solve this problem in the near future.

One of the most promising alternatives to computer-aided design is to follow nature's way of adaptively designing molecules, i.e. by molecular evolution. We call the application of molecular evolution for designing molecules "evolutionary biotechnology" (Eigen and Rigler 1994; Schuster 1995; Koltermann and Kettling 1997). Evolutionary biotechnology mimics the basic steps of natural evolution in a way that the outcome is a molecular variant with the intended phenotype. While biotechnology's first phase was characterized by the exploration and the use of nature's molecules for human purposes, the second phase is now going to be characterized by the use of nature's means of creating new molecules.

## 3 How Does Nature Create Functional Molecules?

"Nothing in biology makes sense except in the light of evolution" (Dobzhansky 1973). This 25-year-old phrase by Dobzhansky underlined a new era in understanding the principles of living matter at that time. Every biological phenomenon has to be interpreted as a result of Darwinian selection over many generations. To elucidate the fundamental principles of why and how Darwinian selection is appropriate for creating molecules with particular functions, we have to analyze the elementary steps of evolution on the molecular level. This task – namely to base

Darwinian evolution on solid physical principles – was initiated in the early 1970s with a quantitative description of nucleic acid replication kinetics under competitive conditions (EIGEN 1971), and since then it has been successively refined (EIGEN and SCHUSTER 1977; EIGEN et al. 1988; SCHUSTER 1995; for a recent detailed review see SCHUSTER 1997). Here, we will emphasize those aspects of the theory that are relevant for evolutionary biotechnology.

First we shall ask: What kind of variable is required for a quantitative description of evolutionary processes? The answer has already been given above, i.e. the fundamental variable of evolution is information. In biology, information means the fixation of a certain structure or sequence that leads reproducibly to a defined three-dimensional structure. The basic elements of biological information are genes, and the universal biological information carriers are nucleic acids, as already mentioned. Since Watson and Crick discovered the double-helix structure of DNA in the 1950s (WATSON and CRICK 1953), we have learned much about the molecular basis of the remarkable properties that make nucleic acids ideal information carriers. It is quite instructive to recall some of them.

Nucleic acids are linear copolymers with an enormous capacity for storing information. With their four different monomers, the storage capacity increases exponentially with chain length – and even at modest lengths an enormous complexity is reached. The large storage capacity is accompanied by only a minor interaction between the macromolecular structure and the information content. In particular for the double-stranded DNA molecule, the composition and the order of monomers has an almost negligible impact on the ability to process and reproduce it. This independence seems to be one of the main prerequisites for any evolutionary creation of complex functions. Furthermore, nucleic acids are able to reproduce via replication in an autocatalytic process. In vivo as well as in vitro, an enzyme (a polymerase) is required to catalyze the template-directed polymerization of nucleic acids. However, the actual task, i.e. making a copy of a nucleic acid, is always carried out by this nucleic acid itself. It is widely accepted that in the early days of molecular evolution on earth this process worked entirely independently (ORGEL 1986). Finally, this replication process is highly accurate, but only to a certain degree. Mutations happen spontaneously during amplification, and once occurred they are inherited infinitely.

We have learned a lot about the basics of molecular evolution from in vitro studies with nucleic acids. These experiments were initiated in the late 1960s by Spiegelman and coworkers, who for the first time studied in vitro evolution by using the Qβ replicase system (MILLS et al. 1967). Subsequent related studies (OEHLENSCHLÄGER and EIGEN 1997; STRUNK and EDERHOF 1997; BIEBRICHER and GARDINER 1997), in combination with theoretical calculations and model developments, revealed some peculiar features of molecular evolution. First, we had to re-define our understanding of the target of evolution. According to the quasispecies concept, the target of selection is not a single mutant but rather a distribution of related mutants occupying a distinct region in sequence space (EIGEN et al. 1988). The expansion of this so-called quasispecies is given by the error rate of the amplification process; and the population density at each point is determined by a combination of the corresponding fitness value and the density at those positions that are connected through mutation events. In contrast to earlier conceptions, which implied completely spontaneous emergence of mutants, a dynamic quasispecies already includes a large spectrum of individual genotypes with different fitness values. Whenever selection pressure changes, it is likely that better adapted variants are already present. Thus, selection is by no means a random drift but a directed shift of the quasispecies distribution towards regions of higher fitness, guided by the underlying fitness landscape. Another important point regards the relationship between sequence space and function. Using RNA as a model, SCHUSTER (1995) showed that: (1) a particular function is coded by several sequences equally distributed over the sequence space, and (2) all common functions are coded within a relatively small radius around a particular sequence. Both characteristics are important aspects for the application of evolutionary biotechnology.

Finally we want to ask how the different basic features of molecular evolution interact to obtain molecules that are obviously optimally adapted to their particular task in the biological context. Molecular evolution is a rather complex interaction of the two simple processes variation and selection. Additional processes are amplification and expression of function (Fig. 1). Under natural circumstances, however, amplification is intrinsically connected to selection and variation; and expression is only required when genotype and phenotype are different molecules. In nature, variation usually derives from spontaneous mutation events, the rate of which is a characteristic constant of each amplification system. It primarily determines adaptation velocity by expanding the quasispecies distribution. However, mathematical analysis demonstrates that beyond a certain threshold of mutation frequency the information content of the quasispecies is lost (SCHUSTER 1986). This error threshold is usually inversely related to the sequence length. Consequently, the highest rate of evolutionary adaptation is found at an error rate slightly below the error threshold. A variety of techniques has been described for artificially introducing mutations and recombination events, e.g. cassette muta-

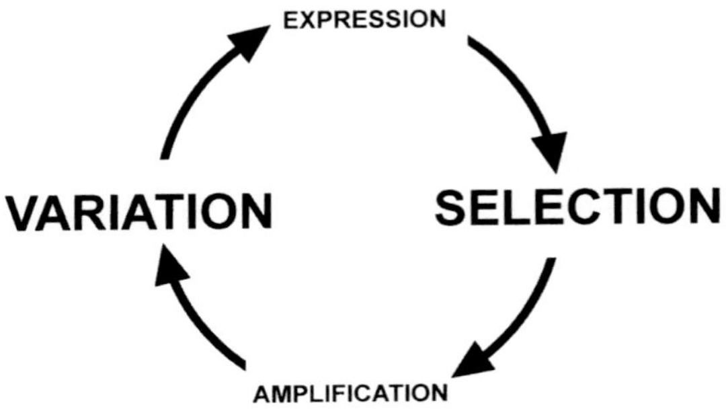


**Fig. 1.** Evolution on the molecular level can be described as the cyclic interaction of variation and selection. In nature, these two processes are inherently connected to amplification; for evolutionary biotechnology, all steps may be separated by technical means. Expression of function is only required when phenotype and genotype carriers are separate molecules

genesis (Dube and Loeb 1989), error-prone PCR (Cadwell and Joyce 1992), or DNA shuffling (Stemmer 1994). Selection, on the other hand, determines the direction of evolution. In natural evolving systems, selection is usually connected to the amplification rate, as described in more detail below. Evolutionary biotechnology, in contrast, enables us to apply alternative selection strategies. However, the mode of selection remains the crucial variable.

## 4 Directing Molecular Evolution by Technical Means: Selection Strategies

Evolutionary biotechnology aims at the creation of genetic information coding for a biomolecule that bears a certain phenotype. Here, we shall define the term molecular phenotype more precisely. Phenotype usually refers to one primary function, possibly accompanied by a set of secondary functions and properties. Common primary molecular functions can be classified as follows: (1) recognition and binding (e.g. nucleic acid aptamers, peptide binders, antibodies), (2) recognition and effecting (e.g. small molecules acting as enzyme inhibitors, peptide hormones), or (3) recognition and catalysis (e.g. enzymes, ribozymes, catalytic antibodies), all of which may be modulated with respect to activity, affinity, and specificity. In addition, there is a variety of more or less complex secondary properties, such as stability under various environmental conditions (temperature, solvent, etc.), size and composition, regulatability, cofactor requirements, or immunogeneity.

Of all the steps of molecular evolution it is the selection step that controls the molecular phenotype (Fig. 1). Selection quantitatively judges the fitness of the molecular phenotype and enriches variants with fitness values above the average of the applied ensemble. Under conditions of natural molecular evolution, selection is intrinsically coupled to the amplification step; due to a direct correlation between the fitness value and the net amplification rate, variants with fitness values above the average amplify exponentially at the expense of all others. This, theoretically, results in a population of neutral mutants, i.e. variants with equally high fitness values (Eigen 1971). The amplification-coupled strategy was successfully realized in directed evolution experiments, e.g. for improving the catalytic activity of enzymes necessary for utilizing the offered carbon source (Hall and Zuzel 1980), for selection of ribozymes with DNA-cleaving activity (Beaudry and Joyce 1992, Kurz and Breaker, this volume), or for designing phage-displayed binders by coupling an antibody-antigen interaction to the infection process (Duenas and Borrebaeck 1994; Krebber et al. 1995). A similar approach has recently been demonstrated by coupling enzyme activity to phage infectivity (Gao et al. 1997). Furthermore, instead of promoting directly the amplification of the system, a net amplification may alternatively be achieved by hindering its destruction in connection with the intended function. Examples for applying this strategy are the selection of enzymes that render bacteria resistant against antibiotics (DUBE and LOEB 1989), or of RNA molecules that are resistant to ribonuclease treatment (STRUNK and EDERHOF 1997). The reverse two-hybrid system may serve as an interesting example of an indirect coupling of function and survival. In contrast to conventional two-hybrid reporter systems, here, the inhibition of a certain protein–protein interaction prevents the induction of a toxic gene promoter (LEANNA and HANNINK 1996; VIDAL et al. 1996; Kolanus, this volume). However, the applicability of systems with amplification-coupled selection is rather limited. Most molecular phenotypes are not suited to be linked directly to the growth rate of the expression host. The complexity of other phenotypes only allows a very indirect linkage and therefore indirect selection. The expression host may find alternative ways to amplify without having realized the originally intended function.

The more generic alternative implies uncoupling of amplification and selection steps. It makes use of selection techniques that are based on direct judging of the molecular phenotype. Enrichment is achieved by artificially isolating all variants with fitness values above an arbitrary threshold, followed by their re-amplification. Ideally, this amplification step is nonselective and treats all variants equally without any influence of their phenotype. In practice, however, amplification-biased effects are common. There are usually some protein variants that are preferred by an expression host and some that cannot be expressed at all (IANNOLO et al. 1997). Even with in vitro selection of RNA molecules, bias effects due to preferences and aversions of the polymerase enzyme have been observed (ELLINGTON and SZOSTAK 1990). Furthermore, there is often a tendency of removing dispensable genetic information. Due to an inverse relation between replication rate and sequence length, exponential amplification favors strongly variants with a shorter genome, as analyzed in detail for RNA in vitro amplification (OEHLENSCHLÄGER and EIGEN 1997).

Amplification-uncoupled selection may be classified as selection by physical separation and selection by screening. Physical separation usually means the extraction of all molecules that bear the intended phenotype in a single step. Such a strategy is the method of choice for selecting binders. Upon binding to a solid support, unspecific binders are separated from specific binders by repeated washing steps. This has been successfully applied to nucleic acids (ELLINGTON and SZOSTAK 1990; TUERK and GOLD 1990; Famulok and Mayer, this volume), peptide libraries (Uebel et al., this volume; Reineke et al., this volume), and small molecules (Eliseev and Lehn, this volume). In the latter case, the necessary physical connection between genotype and phenotype is usually met by any display technique, e.g. phage display (SMITH 1985; Ge and Johnsson, this volume), bacterial display (FUCHS et al. 1991), or polysome display and related RNA-peptide fusions (MATTHEAKIS et al. 1994; HANES and PLÜCKTHUN 1997; ROBERTS and SZOSTAK 1997; Hanes and Plückthun, this volume). Separation by specific binding to a solid support has also been adapted to the selection of catalysts. Catalytic antibodies have been selected through binding to a transition state analog (TRAMONTANO et al. 1986). More recently, phage-displayed enzymes were selected which modify themselves or their fusion proteins, leading to a modification that can be selected by affinity to an immobilized target (Soumillion et al. 1994; Pedersen et al. 1998). Furthermore, such a selection strategy is not limited to separation on solid supports. For example, the selection of ribozyme molecules of a certain size can be achieved by gel separation (Heidenreich and Eckstein 1997). However, the limits of physical separation processes are reached when we intend to create more complex biochemical functions, such as specific activation of cellular receptors (in contrast to simple binding) or catalysts with tightly controlled specificity and selectivity. The logical alternatives are procedures in which each molecular variant is examined individually (i.e. screened) and – depending on the result – sorted out and selected for re-amplification. Selection by screening implies important characteristics which shall be discussed in more detail.

Selection by screening deals with single variants. Usually, pure clones of isolated variants or defined mixtures of them are subjected to a screening process. A physical linkage between phenotype and genotype is not necessary here. However, clones or mixtures of them have to be spatially separated to guarantee the right assignment of fitness values and information carriers. This can be achieved by simply spreading clones on an agar plate (Zhang et al. 1997) or by any compartmentalization, e.g. in the wells of a sample carrier (Moore and Arnold 1996; Schober et al. 1997), in a capillary (Bauer et al. 1989), or in cells or vesicles (Tawfik and Griffiths 1998). The most important prerequisite for screening is the access to an assay that allows precise characterization of the phenotype. Such an assay, in combination with a suitable detection device, represents the crucial aspect of screening approaches. Its sensitivity determines the minimum of individuals needed in a single sample, which in turn defines the overall consumption of material. Its ability or inability to imitate the actually intended function determines the discrepancy between the resulting and the intended phenotype; lower similarity possibly requires one or more correction steps, as demonstrated in detail for selection by screening of an esterase (Moore and Arnold 1996). Finally, the measurement time per sample as well as the number of manipulations involved in the assay define the throughput rate and therefore limit the size of the population to be screened. However, the number of possible variants of a given average-sized sequence by far exceeds the number that can be handled in reasonable time in the laboratory, i.e. the complexity problem. Therefore, in order to avoid that a selection-by-screening approach gets stuck in a local optimum, the throughput rate has to be kept sufficiently high; depending on the information complexity it should range at least between $10^4$ and $10^6$ screened variants per cycle. Of course, the complexity can generally be extended by combining several means, such as: (1) mix-and-split techniques, (2) continuous adaptation in fitness space through the application of many successive selection rounds (Eigen et al. 1988), and (3) by allowing recombination events in order to recombine individually selected genotypes and to remove accumulated mutations with negative effects (Stemmer 1994; Steipe, this volume). The advantages of selection by screening are obvious. Basically, any function can be selected provided that a suitable assay is accessible. Environmental conditions are not restricted to those required for the survival of cells since it is usually sufficient to isolate and re-amplify the genetic information, e.g. by PCR.

And selection pressure can be designed and modulated at will. Therefore, the final result of screening-based selection is better to control, particularly with respect to the fine tuning of secondary features such as environmental influences or cofactor requirements.

The limited number of variants to be judged per evolutionary cycle and the requirement to clone and amplify these variants are the major obstacles of conventional screening-based selection. As a consequence, fluorescence-based single-molecule detection and manipulation techniques have been developed and will be further optimized to overcome these obstacles (Eigen and Rigler 1994; Koltermann et al. 1998). A discussion of current ideas and perspectives in this field shall conclude this article.

## 5 Confocal Fluorescence Spectroscopy for Evolutionary Biotechnology

Among the physical properties that are sufficiently specific and sensitive for characterizing individual molecules, fluorescence has an outstanding importance. By using highly efficient fluorophore labels and a small illuminated laser focus as the detection volume, a variety of molecular characteristics has been successfully analyzed at the single-molecule level, e.g. translational diffusion constants (Ehrenberg and Rigler 1976), velocity and direction of hydrodynamic flow (Brinkmeier and Rigler 1995), binding kinetics and dissociation constants (Rauer et al. 1996; Schwille et al. 1996), catalytic rate constants (Xue and Yeung 1995; Kettling et al. 1998), or kinetics of conformational changes (Ha et al. 1996; Eggeling et al. 1998).

Conventional fluorescence detection averages fluorescence emission signals over space and time. Resolution of single events is prevented by the fact that many fluorescence photons originating from a large ensemble of molecules are superimposed. Noise from background fluorescence and other interfering radiation further deteriorate the resolution. In order to detect and characterize single molecules it is necessary to optimize spatial and temporal resolution. A high spatial resolution can be reached using confocal optics. Here, the probe volume is restricted to a focal spot of less than one femtoliter (i.e. the size of a typical bacterial cell) by epi-illumination of a microscope objective with appropriate laser beams and by imaging the collected photons onto sensitive single-photon detectors. A sufficiently high temporal resolution (down to the range of nanoseconds) is achieved by using avalanche photo diodes in combination with suitable data collection systems (Fig. 2). With fluorophore concentrations of one nanomole per liter and below, less than one molecule resides on average in the femtoliter focal volume. Therefore, fluorescence bursts can be assigned to those single molecules that enter and leave the probe volume. Detection of single fluorescent molecules is then simply achieved by looking for intensity peaks while scanning the focus through the sample.

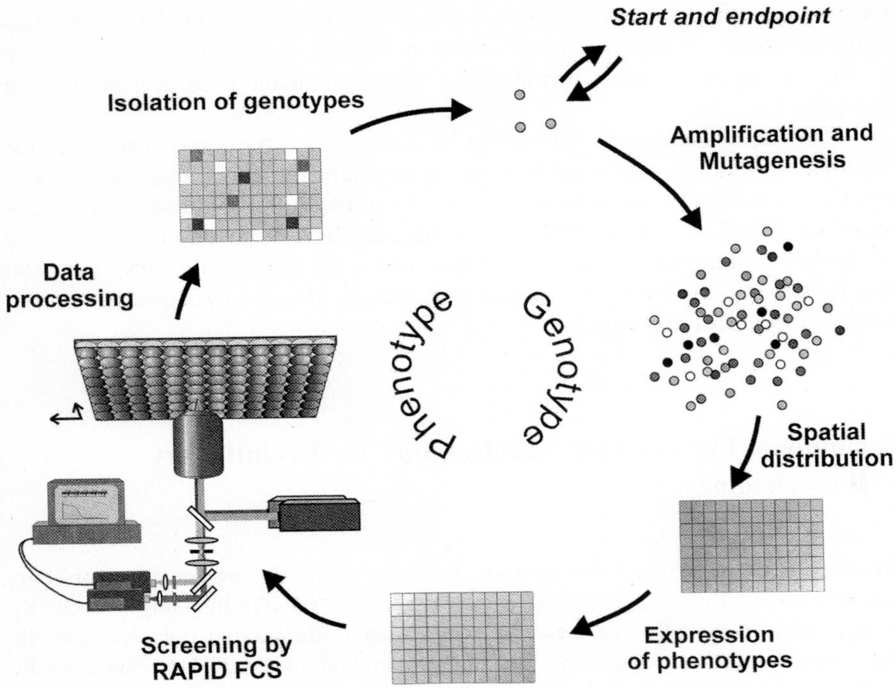


**Fig. 2.** Rapid assay processing by integration of dual-color fluorescence cross-correlation spectroscopy (RAPID FCS)-based evolutionary biotechnology. A library of genotypes is dispensed into compartments of a sample carrier. After expressing the molecular phenotype, individual variants are screened by RAPID FCS, and corresponding genotypes with increased levels of intended functions are selected for the next cycle of variation and selection

However, characterization of molecules in order to distinguish different species within a population requires additional data evaluation. Signals have to be traced with high temporal resolution and evaluated by powerful data processing. For example, differences in molecular mass and hydrodynamic radius can easily be characterized by evaluating rotational or translational diffusion constants by means of fluorescence correlation spectroscopy (FCS) (Eigen and Rigler 1994). Shifts of fluorescence intensity per particle can be analyzed by fluorescence intensity distribution analysis. Likewise, changes in conformation or of the chemical environment of a molecule may be detected by fluorescence life-time spectroscopy or by wavelength-selective spectroscopy in order to detect possible shifts of the decay time or of the emission spectrum.

Recently, conventional single-color fluorescence spectroscopy has efficiently been extended to dual-color fluorescence spectroscopy (Schwille et al. 1997). Dual-color analysis has several advantageous characteristics. For dual-color spectroscopy two pairs of excitation and detection wavelengths are combined. This allows the tracing of two spectrally separated fluorophores at the same time and a combined evaluation of their signals. For example, cross-correlation analysis between both signals gives access to the number of correlated fluorescence fluctuations in different colors and their time constants. Compared with single-color auto-correlation, the employment of two distinct labels provides drastically increased signal specificity and accuracy. Therefore it allows characterization of molecular properties at higher background levels with shorter analysis times per sample. This makes dual-color fluorescence cross-correlation spectroscopy an ideal tool for screening applications in biochemistry and cellular biology (KOLTERMANN et al. 1998). Even at large excess of background fluorescence in each channel, dual-color cross-correlation enables specific detection of double fluorescent molecules. Consequently, the typical assay format is based on distinguishing single-labeled from double-labeled molecules. Any reaction that can be linked to the formation or destruction of a linkage between two fluorophores can easily be examined. Using an endonuclease assay, dual-color cross-correlation spectroscopy has successfully been applied to large number screening, a combination that was termed RAPID FCS (rapid assay processing by integration of dual-color fluorescence cross-correlation spectroscopy) (KOLTERMANN et al. 1998). Data collection times for precise determination of endonucleolytic activity by RAPID FCS lay in the range of ≤1s, which is clearly faster than single-color FCS and corresponds to a screening throughput up to $10^5$ samples per day.

Besides cross-correlation FCS, alternative algorithms for data processing of multi-color fluorescence signals originating from single molecules are currently being examined. Special attention has been given to the extraction of accurate signals at the shortest analysis times possible. Preliminary results of recent efforts revealed sampling times in the range of 100ms, which would allow sampling of approximately $10^6$ variants per day (WINKLER et al. 1999). By combining RAPID FCS and additional detection principles with automated sample processing devices, a universally applicable, automated selection tool is being developed (Fig. 2). The emphasis of these efforts is on the evolutionary optimization of catalysts for cleavage and ligation reactions. Possible applications are changing and improving the activity and substrate specificity of endonucleases, proteases, and related enzymes. The main advantages of the system are the possibility to choose any selection criteria for catalysts and to precisely control velocity and direction of the evolutionary process.

Although based on single-molecule fluctuations, RAPID FCS requires a reasonable number of these events to give accurate values. Therefore, it is still an ensemble of each molecular species to be traced. With respect to selection within an evolutionary optimization, this means that the individual variants to be screened have to be separated spatially, e.g. in compartments of a nanocarrier or on a flat surface. If the molecules to be optimized are catalysts such as enzymes, it is sufficient to isolate individual cells which each express a particular enzyme variant and to detect an ensemble of occasionally converted substrate molecules. However, in contrast to such an indirect characterization, the remaining challenge is the direct selection and sorting of individual variants of a protein according to their molecular properties. Such single-molecule sorting will undoubtedly be the ultimate goal of screening-based approaches for evolutionary design. It may be achieved by a combination of a scanner and a picker device, or by using capillaries or microstructures with integrated sorting modules (EIGEN and RIGLER 1994). Currently, we are studying the application of FCS and other confocal fluorescence techniques in microstructures with different architectures including detection channels, mixing chambers and selection crosses (BRINKMEIER et al. 1997). In this context, our experiences well match our expectations of several years ago. Confocal fluorescence spectroscopy is not only preferably suited for miniaturization due to its high performance and small detection volume, a comprehensive integration into nanotechnology will even increase the benefit from its exceptional features. Their successful combination with mutagenesis techniques makes confocal fluorescence spectroscopy one of the most powerful tools for selection by screening strategies in evolutionary biotechnology.

*Acknowledgements.* The authors wish to thank Petra Schwille and Thorsten Winkler for many stimulating discussions, and Claudia Eggert for critically reading the manuscript.

## References

Bauer GJ, McCaskill JS, Otten H (1989) Traveling waves of in vitro evolving RNA. Proc Natl Acad Sci USA 86:7937–7941

Beaudry AA and Joyce GF (1992) Directed evolution of an RNA enzyme. Science 257:635–641

Biebricher CK and Gardiner WC (1997) Molecular evolution of RNA in vitro. Biophys Chem 66:179–192

Brinkmeier M, Dörre K, Riebeseel K, Rigler R (1997) Confocal spectroscopy in microstructures. Biophys Chem 66:229–239

Brinkmeier M and Rigler R (1995) Flow analysis by means of fluorescence correlation spectroscopy. Experimental Techniques of Physics 41:205–210

Cadwell RC and Joyce GF (1992) Randomization of genes by PCR mutagenesis. PCR Methods Appl 2:28–33

Cohen SN, Chang AC, Boyer HW, Helling RB (1973) Construction of biologically functional bacterial plasmids in vitro. Proc Natl Acad Sci USA 70:3240–3244

Dobzhansky T (1973) Nothing in biology makes sense except in the light of evolution. Am Bio Teacher 35:125–129

Dube DK and Loeb LA (1989) Mutants generated by the insertion of random oligonucleotides into the active site of the beta-lactamase gene. Biochemistry 28:5703–5707

Duenas M and Borrebaeck CA (1994) Clonal selection and amplification of phage displayed antibodies by linking antigen recognition and phage replication. Biotechnology 12:999–1002

Eggeling C, Fries JR, Brand L, Günther R, Seidel CAM (1998) Monitoring conformational dynamics of a single molecule by selective fluorescence spectroscopy. Proc Natl Acad Sci USA 95:1556–1561

Ehrenberg M and Rigler R (1976) Fluorescence correlation spectroscopy applied to rotational diffusion of macromolecules. Quart Rev Biophys 9:69–81

Eigen M (1971) Selforganization of matter and the evolution of biological macromolecules. Naturwissenschaften 58:465–523

Eigen M, McCaskill J, Schuster P (1988) Molecular Quasi-Species. Journal of Physical Chemistry 92:6881–6891

Eigen M and Rigler R (1994) Sorting single molecules: application to diagnostics and evolutionary biotechnology. Proc Natl Acad Sci USA 91:5740–5747

Eigen M and Schuster P (1977) The hypercycle. A principle of natural self-organization. Naturwissenschaften 64:541–565

Ellington AD and Szostak JW (1990) In vitro selection of RNA molecules that bind specific ligands. Nature 346:818–822

Fuchs P, Breitling F, Dubel S, Seehaus T, Little M (1991) Targeting recombinant antibodies to the surface of Escherichia coli: fusion to a peptidoglycan associated lipoprotein. Biotechnology 9:1369–1372
Gao CS, Lin CH, Lo CHL, Mao S, Wirsching P, Lerner RA, Janda KD (1997) Making chemistry selectable by linking it to infectivity. Proc Natl Acad Sci USA 94:11777–11782
Gibbons A (1992) Biotech's Second Generation. Science 256:766–768
Ha T, Enderle T, Ogletree DF, Chemla DS, Selvin PR, Weiss S (1996) Probing the interaction between two single molecules: Fluorescence resonance energy transfer between a single donor and a single acceptor. Proc Natl Acad Sci USA 93:6264–6268
Hall BG and Zuzel T (1980) Evolution of a new enzymatic function by recombination within a gene. Proc Natl Acad Sci USA 77:3529–3533
Hanes J and Plückthun A (1997) In vitro selection and evolution of functional proteins by using ribosome display. Proc Natl Acad Sci USA 94
Heidenreich O and Eckstein F (1997) Synthetic Ribozymes: The Hammerhead Ribozyme. Concepts in Gene Therapy. Eds: Strauss M, Barranger JA. Walter de Gruyter, 169
Iannolo G, Minenkova O, Gonfloni S, Castagnoli L, Cesareni G (1997) Construction, exploitation and evolution of a new peptide library displayed at high-density by fusion to the major coat protein of filamentous phage. Biological Chemistry 378:517–521
Kettling U, Koltermann A, Schwille P, Eigen M (1998) Real-time enzyme kinetics monitored by dual-color fluorescence cross-correlation spectroscopy. Proc Natl Acad Sci USA 95:1416–1420
Koltermann A, Kettling U, Bieschke J, Winkler T, Eigen M (1998) Rapid assay processing by integration of dual-color fluorescence cross-correlation spectroscopy: High throughput screening for enzyme activity. Proc Natl Acad Sci USA 95:1421–1426
Koltermann A and Kettling U (1997) Principles and Methods of Evolutionary Biotechnology. Biophys Chem 66:159–177
Krebber C, Spada S, Desplancq D, Plückthun A (1995) Co-selection of cognate antibody-antigen pairs by selectively-infective phages. FEBS Lett 377:227–231
Leanna CA and Hannink M (1996) The reverse two-hybrid system: a genetic scheme for selection against specific protein–protein interactions. Nucleic Acids Res 24:3341–3347
Mattheakis LC, Bhatt RR, Dower WJ (1994) An in vitro polysome display system for identifying ligands from very large peptide libraries. Proc Natl Acad Sci USA 91:9022–9026
Mills DR, Peterson RL, Spiegelman S (1967) An extracellular Darwinian experiment with a self-duplicating nucleic acid molecule. Proc Natl Acad Sci USA 58:217–224
Moore JC and Arnold FH (1996) Directed evolution of a para-nitrobenzyl esterase for aqueous-organic solvents. Nat Biotechnol 14:458–467
Oehlenschläger F and Eigen M (1997) 30 years later – a new approach to Sol Spiegelmans and Leslie Orgels in vitro evolutionary studies. Orig Life Evol Biosph 27:437–457
Orgel LE (1986) RNA catalysis and the origins of life. J Theor Biol 123:127–149
Pedersen H, Holder S, Sutherlin DP, Schwitter U, King DS, Schultz PG (1998) A method for directed evolution and functional cloning of enzymes. Proc Natl Acad Sci USA 95:10523–10528
Rauer B, Neumann E, Widengren J, Rigler R (1996) Fluorescence correlation spectrometry of the interaction kinetics of tetramethylrhodamine alpha-bungarotoxin with Torpedo californica acetylcholine receptor. Biophys Chem 58:3–12
Roberts RW and Szostak JW (1997) RNA-peptide fusions for the in vitro selection of peptides and proteins. Proc Natl Acad Sci USA 94:12297–12302
Schober A, Günther R, Tangen U, Goldmann G, Ederhof T, Koltermann A, Wienecke A, Schwienhorst A, Eigen M (1997) High throughput screening by multichannel glass fiber fluorimetry. Rev Sci Instrum 68:2187–2194
Schuster P (1986) The physical basis of molecular evolution. Chemica Scripta 26:27–41
Schuster P (1995) How to search for RNA structures – Theoretical concepts in evolutionary biotechnology. J Biotechnol 41:239–257
Schuster P (1997) Genotypes with phenotypes – Adventures in an RNA toy world. Biophys Chem 66: 75–110
Schwille P, Oehlenschläger F, Walter NG (1996) Quantitative hybridization kinetics of DNA probes to RNA in solution followed by diffusional fluorescence correlation analysis. Biochemistry 35:10182–10193
Schwille P, Meyer-Almes FJ, Rigler R (1997) Dual-color fluorescence cross-correlation spectroscopy for multicomponent diffusional analysis in solution. Biophys J 72:1878–1886
Smith GP (1985) Filamentous fusion phage: novel expression vectors that display cloned antigens on the virion surface. Science 228:1315–1317

Soumillion P, Jespers L, Bouchet M, Marchand-Brynaert J, Sartiaux P, Fastrez J (1994) Phage display of enzymes and in vitro selection for catalytic activity. Appl Biochem Biotechnol 47:175–89

Stemmer WP (1994) Rapid evolution of a protein in vitro by DNA shuffling [see comments]. Nature 370:389–391

Strunk G and Ederhof T (1997) Machines for automated evolution experiments in vitro based on the serial-transfer concept. Biophys Chem 66:193–202

Tawfik DS and Griffiths AD (1998) Man-made cell-like compartments for molecular evolution. Nat Biotechnol 16:652–656

Tramontano A, Janda KD, Lerner RA (1986) Catalytic antibodies. Science 234:1566–1570

Tuerk C and Gold L (1990) Systematic evolution of ligands by exponential enrichment: RNA ligands to bacteriophage T4 DNA polymerase. Science 249:505–510

Vidal M, Brachmann RK, Fattaey A, Harlow E, Boeke JD (1996) Reverse two-hybrid and one-hybrid systems to detect dissociation of protein–protein and DNA-protein interactions. Proc Natl Acad Sci USA 93:10315–10320

Watson JD and Crick FH (1953) Molecular structure of nucleic acids: a structure for deoxyribose nucleic acid. Nature 177:964

Winkler T, Kettling U, Koltermann A, Eigen M (1999) Confocal fluorescence coincidence analysis: an approach to ultra high-throughput screening. Proc Natl Acad Sci USA 96:1375–1378

Xue Q and Yeung ES (1995) Differences in the chemical reactivity of individual molecules of an enzyme. Nature 373:681–683

Zhang JH, Dawes G, Stemmer WPC (1997) Directed evolution of a fucosidase from a galactosidase by dna shuffling and screening. Proc Natl Acad Sci USA 94:4504–4509

# Subject Index

# Current Topics in Microbiology and Immunology

## Volumes published since 1989 (and still available)

Vol. 201: **Kosco-Vilbois, Marie H. (Ed.):** An Antigen Depository of the Immune System: Follicular Dendritic Cells. 1995. 39 figs. IX, 209 pp. ISBN 3-540-59013-7

Vol. 202: **Oldstone, Michael B. A.; Vitković, Ljubiša (Eds.):** HIV and Dementia. 1995. 40 figs. XIII, 279 pp. ISBN 3-540-59117-6

Vol. 203: **Sarnow, Peter (Ed.):** Cap-Independent Translation. 1995. 31 figs. XI, 183 pp. ISBN 3-540-59121-4

Vol. 204: **Saedler, Heinz; Gierl, Alfons (Eds.):** Transposable Elements. 1995. 42 figs. IX, 234 pp. ISBN 3-540-59342-X

Vol. 205: **Littman, Dan R. (Ed.):** The CD4 Molecule. 1995. 29 figs. XIII, 182 pp. ISBN 3-540-59344-6

Vol. 206: **Chisari, Francis V.; Oldstone, Michael B. A. (Eds.):** Transgenic Models of Human Viral and Immunological Disease. 1995. 53 figs. XI, 345 pp. ISBN 3-540-59341-1

Vol. 207: **Prusiner, Stanley B. (Ed.):** Prions Prions Prions. 1995. 42 figs. VII, 163 pp. ISBN 3-540-59343-8

Vol. 208: **Farnham, Peggy J. (Ed.):** Transcriptional Control of Cell Growth. 1995. 17 figs. IX, 141 pp. ISBN 3-540-60113-9

Vol. 209: **Miller, Virginia L. (Ed.):** Bacterial Invasiveness. 1996. 16 figs. IX, 115 pp. ISBN 3-540-60065-5

Vol. 210: **Potter, Michael; Rose, Noel R. (Eds.):** Immunology of Silicones. 1996. 136 figs. XX, 430 pp. ISBN 3-540-60272-0

Vol. 211: **Wolff, Linda; Perkins, Archibald S. (Eds.):** Molecular Aspects of Myeloid Stem Cell Development. 1996. 98 figs. XIV, 298 pp. ISBN 3-540-60414-6

Vol. 212: **Vainio, Olli; Imhof, Beat A. (Eds.):** Immunology and Developmental Biology of the Chicken. 1996. 43 figs. IX, 281 pp. ISBN 3-540-60585-1

Vol. 213/I: **Günthert, Ursula; Birchmeier, Walter (Eds.):** Attempts to Understand Metastasis Formation I. 1996. 35 figs. XV, 293 pp. ISBN 3-540-60680-7

Vol. 213/II: **Günthert, Ursula; Birchmeier, Walter (Eds.):** Attempts to Understand Metastasis Formation II. 1996. 33 figs. XV, 288 pp. ISBN 3-540-60681-5

Vol. 213/III: **Günthert, Ursula; Schlag, Peter M.; Birchmeier, Walter (Eds.):** Attempts to Understand Metastasis Formation III. 1996. 14 figs. XV, 262 pp. ISBN 3-540-60682-3

Vol. 214: **Kräusslich, Hans-Georg (Ed.):** Morphogenesis and Maturation of Retroviruses. 1996. 34 figs. XI, 344 pp. ISBN 3-540-60928-8

Vol. 215: **Shinnick, Thomas M. (Ed.):** Tuberculosis. 1996. 46 figs. XI, 307 pp. ISBN 3-540-60985-7

Vol. 216: **Rietschel, Ernst Th.; Wagner, Hermann (Eds.):** Pathology of Septic Shock. 1996. 34 figs. X, 321 pp. ISBN 3-540-61026-X

Vol. 217: **Jessberger, Rolf; Lieber, Michael R. (Eds.):** Molecular Analysis of DNA Rearrangements in the Immune System. 1996. 43 figs. IX, 224 pp. ISBN 3-540-61037-5

Vol. 218: **Berns, Kenneth I.; Giraud, Catherine (Eds.):** Adeno-Associated Virus (AAV) Vectors in Gene Therapy. 1996. 38 figs. IX,173 pp. ISBN 3-540-61076-6

Vol. 219: **Gross, Uwe (Ed.):** Toxoplasma gondii. 1996. 31 figs. XI, 274 pp. ISBN 3-540-61300-5

Vol. 220: **Rauscher, Frank J. III; Vogt, Peter K. (Eds.):** Chromosomal Translocations and Oncogenic Transcription Factors. 1997. 28 figs. XI, 166 pp. ISBN 3-540-61402-8

Vol. 221: **Kastan, Michael B. (Ed.):** Genetic Instability and Tumorigenesis. 1997. 12 figs.VII, 180 pp. ISBN 3-540-61518-0

Vol. 222: **Olding, Lars B. (Ed.):** Reproductive Immunology. 1997. 17 figs. XII, 219 pp. ISBN 3-540-61888-0

Vol. 223: **Tracy, S.; Chapman, N. M.; Mahy, B. W. J. (Eds.):** The Coxsackie B Viruses. 1997. 37 figs. VIII, 336 pp. ISBN 3-540-62390-6

Vol. 224: **Potter, Michael; Melchers, Fritz (Eds.):** C-Myc in B-Cell Neoplasia. 1997. 94 figs. XII, 291 pp. ISBN 3-540-62892-4

Vol. 225: **Vogt, Peter K.; Mahan, Michael J. (Eds.):** Bacterial Infection: Close Encounters at the Host Pathogen Interface. 1998. 15 figs. IX, 169 pp. ISBN 3-540-63260-3

Vol. 226: **Koprowski, Hilary; Weiner, David B. (Eds.):** DNA Vaccination/Genetic Vaccination. 1998. 31 figs. XVIII, 198 pp. ISBN 3-540-63392-8

Vol. 227: **Vogt, Peter K.; Reed, Steven I. (Eds.):** Cyclin Dependent Kinase (CDK) Inhibitors. 1998. 15 figs. XII, 169 pp. ISBN 3-540-63429-0

Vol. 228: **Pawson, Anthony I. (Ed.):** Protein Modules in Signal Transduction. 1998. 42 figs. IX, 368 pp. ISBN 3-540-63396-0

Vol. 229: **Kelsoe, Garnett; Flajnik, Martin (Eds.):** Somatic Diversification of Immune Responses. 1998. 38 figs. IX, 221 pp. ISBN 3-540-63608-0

Vol. 230: **Kärre, Klas; Colonna, Marco (Eds.):** Specificity, Function, and Development of NK Cells. 1998. 22 figs. IX, 248 pp. ISBN 3-540-63941-1

Vol. 231: **Holzmann, Bernhard; Wagner, Hermann (Eds.):** Leukocyte Integrins in the Immune System and Malignant Disease. 1998. 40 figs. XIII, 189 pp. ISBN 3-540-63609-9

Vol. 232: **Whitton, J. Lindsay (Ed.):** Antigen Presentation. 1998. 11 figs. IX, 244 pp. ISBN 3-540-63813-X

Vol. 233/I: **Tyler, Kenneth L.; Oldstone, Michael B. A. (Eds.):** Reoviruses I. 1998. 29 figs. XVIII, 223 pp. ISBN 3-540-63946-2

Vol. 233/II: **Tyler, Kenneth L.; Oldstone, Michael B. A. (Eds.):** Reoviruses II. 1998. 45 figs. XVI, 187 pp. ISBN 3-540-63947-0

Vol. 234: **Frankel, Arthur E. (Ed.):** Clinical Applications of Immunotoxins. 1999. 16 figs. IX, 122 pp. ISBN 3-540-64097-5

Vol. 235: **Klenk, Hans-Dieter (Ed.):** Marburg and Ebola Viruses. 1999. 34 figs. XI, 225 pp. ISBN 3-540-64729-5

Vol. 236: **Kraehenbuhl, Jean-Pierre; Neutra, Marian R. (Eds.):** Defense of Mucosal Surfaces: Pathogenesis, Immunity and Vaccines. 1999. 30 figs. IX, 296 pp. ISBN 3-540-64730-9

Vol. 237: **Claesson-Welsh, Lena (Ed.):** Vascular Growth Factors and Angiogenesis. 1999. 36 figs. X, 189 pp. ISBN 3-540-64731-7

Vol. 238: **Coffman, Robert L.; Romagnani, Sergio (Eds.):** Redirection of Th1 and Th2 Responses. 1999. 6 figs. IX, 148 pp. ISBN 3-540-65048-2

Vol. 239: **Vogt, Peter K.; Jackson, Andrew O. (Eds.):** Satellites and Defective Viral RNAs. 1999. 39 figs. XVI, 179 pp. ISBN 3-540-65049-0

Vol. 240: **Hammond, John; McGarvey, Peter; Yusibov, Vidadi (Eds.):** Plant Biotechnology. 1999. 12 figs. XII, 196 pp. ISBN 3-540-65104-7

Vol. 241: **Westblom, Tore U.; Czinn, Steven J.; Nedrud, John G. (Eds.):** Gastroduodenal Disease and Helicobacter pylori. 1999. 35 figs. XI, 313 pp. ISBN 3-540-65084-9

Vol. 242: **Hagedorn, Curt H.; Rice, Charles M. (Eds.):** The Hepatitis C Viruses. 1999. 47 figs. approx. IX, 380 pp. ISBN 3-540-65358-9

Printing: Saladruck, Berlin
Binding: H. Stürtz AG, Würzburg